T0309412

Problem-Solving Strategies in

Mathematics

From Common Approaches to Exemplary Strategies

Problem Solving in Mathematics and Beyond

Series Editor: Alfred S. Posamentier *(Mercy College New York, USA)*

Vol. 1 Problem-Solving Strategies in Mathematics:
From Common Approaches to Exemplary Strategies
by Alfred S. Posamentier and Stephen Krulik

Problem Solving in Mathematics and Beyond
Volume 01

Problem-Solving Strategies in Mathematics

From Common Approaches to Exemplary Strategies

Alfred S. Posamentier
Mercy College New York, USA

Stephen Krulik
Temple University, USA

NEW JERSEY • LONDON • SINGAPORE • BEIJING • SHANGHAI • HONG KONG • TAIPEI • CHENNAI

Published by

World Scientific Publishing Co. Pte. Ltd.
5 Toh Tuck Link, Singapore 596224
USA office: 27 Warren Street, Suite 401-402, Hackensack, NJ 07601
UK office: 57 Shelton Street, Covent Garden, London WC2H 9HE

Library of Congress Cataloging-in-Publication Data
Posamentier, Alfred S.
Problem-solving strategies in mathematics : from common approaches to exemplary strategies / Alfred S. Posamentier (Mercy College, New York, USA) & Stephen Krulik (Temple University, Philadelphia, PA, USA).
pages cm. -- (Problem solving in mathematics and beyond ; vol. 1)
ISBN 978-9814651622 (hc) -- ISBN 978-9814651639 (pbk)
1. Problem solving--Study and teaching (Elementary) 2. Mathematics--Study and teaching (Elementary) I. Krulik, Stephen. II. Title.
QA135.6.P648 2015
372.7--dc23

2015001010

British Library Cataloguing-in-Publication Data
A catalogue record for this book is available from the British Library.

In-House Editors: Tan Rok Ting/V. Vishnu Mohan

Typeset by Stallion Press
Email: enquiries@stallionpress.com

Printed in Singapore

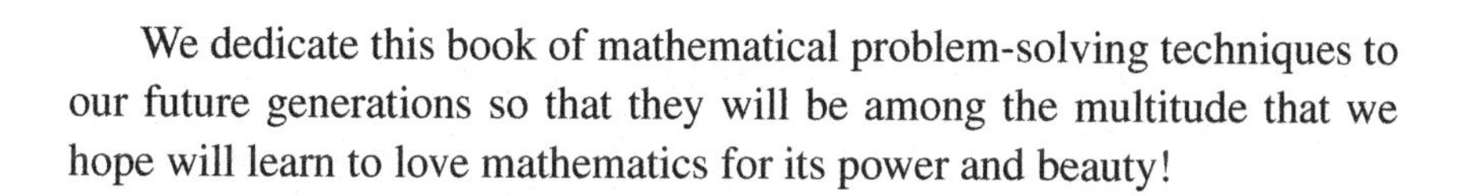

We dedicate this book of mathematical problem-solving techniques to our future generations so that they will be among the multitude that we hope will learn to love mathematics for its power and beauty!

To our children and grandchildren, whose future is unbounded

Lisa, Daniel, David, Lauren, Max, Samuel, and Jack

— Alfred S. Posamentier

Nancy, Dan, Jeff, Amy, Amanda, Ian, Sarah, and Emily

— Stephen Krulik

About the Authors

Alfred S. Posamentier is currently Dean of the School of Education and Professor of Mathematics Education at Mercy College, New York, and previously Distinguished Lecturer at New York City College of Technology of the City University of New York. He is Professor Emeritus of Mathematics Education at The City College of the City University of New York, and former Dean of the School of Education, where he was for 40 years. He is the author and co-author of more than 55 mathematics books for teachers, secondary and elementary school students, and the general readership. Dr. Posamentier is also a frequent commentator in newspapers and journals on topics relating to education.

After completing his B.A. degree in mathematics at Hunter College of the City University of New York, he took a position as a teacher of mathematics at Theodore Roosevelt High School (Bronx, New York), where he focused his attention on improving the students' problem-solving skills and at the same time enriching their instruction far beyond what the traditional textbooks offered. During his six-year tenure there, he also developed the school's first mathematics teams (both at the junior and senior level). He is still involved in working with mathematics teachers and supervisors, nationally and internationally, to help them maximize their effectiveness.

Immediately upon joining the faculty of the City College in 1970 (after having received his master's degree there in 1966), he began to develop inservice courses for secondary school mathematics teachers, including such special areas as recreational mathematics and problem solving in mathematics. As Dean of the City College School of Education for 10 years, his scope of interest in educational issues covered the full gamut educational issues. During his tenure as dean he took the School from the bottom of the New York State rankings to the top with a perfect NCATE accreditation assessment in 2009.

In 2014, Dr. Posamentier repeated this fine achievement by taking the Mercy College School of Education to the unique status of being the only college in the United States to have achieved a perfect initial accreditation rating from both NCATE and CAEP simultaneously.

In 1973, Dr. Posamentier received his Ph.D. from Fordham University (New York) in mathematics education and has since extended his reputation in mathematics education to Europe. He has been visiting professor at several European universities in Austria, England, Germany, Czech Republic, and Poland, while at the University of Vienna he was Fulbright Professor (1990).

In 1989 he was awarded an *Honorary Fellow* at the South Bank University (London, England). In recognition of his outstanding teaching, the City College Alumni Association named him *Educator of the Year* in 1994, and in 2009. New York City had the *day*, May 1, 1994, named in his honor by the President of the New York City Council. In 1994, he was also awarded the *Grand Medal of Honor* from the Republic of Austria, and in 1999, upon approval of Parliament, the President of the Republic of Austria awarded him the title of *University Professor of Austria*. In 2003 he was awarded the title of *Ehrenbürger* (Honorary Fellow) of the Vienna University of Technology, and in 2004 was awarded the *Austrian Cross of Honor for Arts and Science, First Class* from the President of the Republic of Austria. In 2005 he was inducted into the *Hunter College Alumni Hall of Fame*, and in 2006 he was awarded the prestigious *Townsend Harris Medal* by the City College Alumni Association. He was inducted into the New York State *Mathematics Educator's Hall of Fame* in 2009, and in 2010 he was awarded the coveted *Christian-Peter-Beuth Prize* in Berlin.

He has taken on numerous important leadership positions in mathematics education locally. He was a member of the New York State Education

Commissioner's Blue Ribbon Panel on the Math-A Regents Exams, and the Commissioner's Mathematics Standards Committee, which redefined the Standards for New York State, and he also served on the New York City schools' Chancellor's Math Advisory Panel.

Dr. Posamentier is a leading commentator on educational issues and continues his long time passion of seeking ways to make mathematics interesting to both teachers, students and the general public — as can be seen from some of his more recent books: *Numbers: Their Tales, Types and Treasures* (Prometheus, 2015), *Teaching Secondary Mathematics: Techniques and Enrichment Units*, 9th Ed. (Pearson, 2015), *Mathematical Curiosities: A Treasure Trove of Unexpected Entertainments* (Prometheus, 2014), *Geometry: Its Elements and Structure* (Dover, 2014), *Magnificent Mistakes in Mathematics* (Prometheus Books, 2013), *100 Commonly Asked Questions in Math Class: Answers that Promote Mathematical Understanding, Grades 6–12* (Corwin, 2013), *What successful Math Teacher Do: Grades 6–12* (Corwin, 2006, 2013),*The Secrets of Triangles: A Mathematical Journey* (Prometheus Books, 2012), *The Glorious Golden Ratio* (Prometheus Books, 2012), *The Art of Motivating Students for Mathematics Instruction* (McGraw-Hill, 2011), *The Pythagorean Theorem: Its Power and Glory* (Prometheus, 2010), *Mathematical Amazements and Surprises: Fascinating Figures and Noteworthy Numbers* (Prometheus, 2009), *Problem Solving in Mathematics: Grades 3–6: Powerful Strategies to Deepen Understanding* (Corwin, 2009), *Problem-Solving Strategies for Efficient and Elegant Solutions, Grades 6–12* (Corwin, 2008), *The Fabulous Fibonacci Numbers* (Prometheus Books, 2007), *Progress in Mathematics*, K-9 textbook series (Sadlier-Oxford, 2006–2009), *What Successful Math Teacher Do: Grades K-5* (Corwin, 2007), *Exemplary Practices for Secondary Math Teachers* (ASCD, 2007), *101+ Great Ideas to Introduce Key Concepts in Mathematics* (Corwin, 2006), π, *A Biography of the World's Most Mysterious Number* (Prometheus Books, 2004), *Math Wonders: To Inspire Teachers and Students* (ASCD, 2003), and *Math Charmers: Tantalizing Tidbits for the Mind* (Prometheus Books, 2003).

Stephen Krulik is Professor Emeritus of Mathematics Education at Temple University in Philadelphia. While at Temple University Dr. Krulik was responsible for the undergraduate and graduate preparation of mathematics teachers for grades K-12, as well as the inservice alertness training of mathematics teachers at the graduate level. He teaches a wide variety of courses, among them, the History of Mathematics, Methods of Teaching Mathematics, and the Teaching of Problem Solving. This latter course grew out of his interest in problem solving and reasoning in the mathematics classroom. His concern that students understand the beauty and value of problem solving as well as the ability to reason led to his interest in problem solving.

Dr. Krulik received his B.A. degree in mathematics from Brooklyn College of the City University of New York, and his M.A. and Ed.D. in mathematics education from Columbia University's Teachers College. Before coming to Temple University, he taught mathematics in the New York City public schools for 15 years. At Lafayette High School in Brooklyn, he created and implemented several courses deigned to prepare students for the SAT examination, while stressing the art of problem solving as opposed to rote memory of algorithms.

Nationally, Dr. Krulik has served as a member of the committee responsible for preparing the *Professional Standards for Teaching Mathematics* of the National Council of Teachers of Mathematics. He was also the editor of the NCTM's 1980 Yearbook *Problem Solving in School Mathematics.* Regionally, he served as president of the Association of Mathematics Teachers of New Jersey, was a member of the editorial team that produced the 1993 publication, *The New Jersey Calculator Handbook*, and was the Editor for their 1997 monograph, *Tomorrow's Lessons*.

His major areas of interest are the teaching of problem solving and reasoning, materials for teaching mathematics, as well as comprehensive assessment in mathematics. He is the author and co-author of more than 30 books for teachers of mathematics, including the *Roads to Reasoning* (grades 1–8) and *Problem Driven Math* (grades 3–8). Dr. Krulik is also the senior problem-solving author for a basal textbook series. Dr. Krulik is a frequent contributor to the professional journals, in mathematics education.

He has served as a consultant to, and has conducted many workshops for, school districts throughout the United States and Canada, as well as delivering major presentations in Vienna (Austria), Budapest (Hungary), Adelaide (Australia), and San Juan (Puerto Rico). He is in great demand as a speaker at both national and international professional meetings, where his major focus is on preparing *all* students to reason and problem solve in their mathematics classroom, as well as in life.

In 2007, he was given the Great Teacher Award by Temple University. In 2011 he was presented with the Lifetime Achievement Award for Distinguished Service to Mathematics Education by the National Council of Teachers of Mathematics.

Contents

Introduction

Ever since the early 1980's, problem solving, reasoning, and critical thinking have been a major thrust of the school mathematics curriculum throughout the United States, and subsequently much of the world. In fact, as early as 1977 the National Council of Supervisors of Mathematics stated that "learning to solve problems is the principal reason for studying mathematics". After all, of what use is knowing how to do something (algebra?) if one does not know *when* to do it. The problem-solving movement has been gathering steam, and growing to encompass a large part of the study of mathematics. As it continues to grow, it has carried over to solving problems in everyday life. Every day, people are confronted with problems to solve. These can range from the very simple, such as what to wear today, to the more complex. Even what *appears* to be a simple problem, such as crossing the street, can become more complex and require definite thought when we move from country to country where cars travel on different sides of the road.

Before we can begin to talk about problem solving, we should first decide on what constitutes a problem. A problem is a situation that confronts the individual that requires resolution, and for which no path to the answer is readily known. Notice the phrase "for which no path to the answer is readily known". After all, when many of us went to school in the United States, the problems we were taught to solve were often "typed". That is, "age problems" were solved by one procedure, "motion problems" solved by still another procedure, "mixture problems", "liquid measure problems", and so on — each solved by one particular method. In fact, once we had learned the appropriate method, these were really not even problems in the true problem-solving sense. All one had to do was to recognize the particular type of problem, and apply the appropriate automatic process.

The history of mathematical achievement is filled with breakthroughs that often elicited the reaction, "I would never have thought of that approach." Even today, when a clever or elegant solution to a problem is presented, many people have the same reaction. Problem-solving attempts to make these unusual solutions a part of an attainable problem-solving knowledge base.

Problem solving today is largely based on the heuristic model developed by George Polya in his book, "How to Solve It", which was published in 1945, and is still available today. In this book, Polya presented the following four-step plan to solving a problem:

(1) Understanding the problem
(2) Devising a plan
(3) Carrying out the plan
(4) Looking back

Most current problem-solving models are based on this four-step heuristic model. The plan usually includes: (1) Read the problem, (2) Select an appropriate strategy, (3) Solve the problem, and (4) Look back or reflect on the solution. The terminology may be different, but the ideas are the same. The key to the entire process is selecting a proper strategy, or deciding how to attack the problem. It is to examine this critical step in detail that this book was developed and written.

As we have said, selecting the appropriate strategy is the key. Many different sets of strategies have been written about, and presented by different authors over the past decades. Most have common threads running throughout them. In this book, we have decided to examine what we consider to be the 10 most valuable strategies to use when solving problems. We have devoted a full chapter to each strategy. In the presentation of problems, we have tried first to suggest what would usually be the most obvious or common approach. Much of the time this approach would lead to a correct answer. However, the most common approach often requires a great deal of confusing algebra, some difficult computation, and sometimes may not even result in the correct answer.

Next, we have suggested a more elegant, or exemplary solution, demonstrating how the problem-solving strategy under consideration will lead to the answer. Notice that we are differentiating between the "answer" and the "solution". The solution is the entire process from the moment we begin to read the problem, until the final answer has been arrived at and reflected upon. It has been said by some people that the actual answer is one of the least important parts of the solution. Yes, it must be correct, but the process by which the answer was arrived at is the crucial part of the solution.

As you read through the book (and, we hope, work through the problems), notice that in many cases it is possible to use more than one strategy to solve the problem. For example, solving a problem using the intelligent-guess-and-test strategy usually requires organizing the data in a neat, orderly manner. When this happens, we have placed the problem in what we consider to be the more appropriate chapter.

In this book, we begin each chapter with a description of the particular strategy, show how it can be applied to some everyday situations, and then present examples of how it can be applied in a mathematics setting. We then present a series of problems that can best be resolved by the particular strategy. Each problem is an attempt to illustrate the use of that particular strategy. The strategies to be considered are as follows:

1. Logical Reasoning
2. Pattern Recognition
3. Working Backwards
4. Adopting a Different Point of View
5. Considering Extreme Cases
6. Solving a Simpler Analogous Problem
7. Organizing Data
8. Making a Drawing or Visual Representation
9. Accounting for All Possibilities
10. Intelligent Guessing and Testing

As we have mentioned earlier, there is rarely one unique way to solve a problem. The solution we have shown is what we consider to be one example of an exemplary solution, but far from unique. We would encourage the reader to try to find other solutions that may be interesting and unusual. If you find other interesting solutions, we say, "Bravo!" Furthermore, there will be times when more than one single strategy may be used in combination with others providing a varying degree of efficiency.

To demonstrate how a problem can be approached (and solved) with a variety of strategies, we offer several solutions to a popular problem.

PROBLEM

In a room with 10 people, everyone shakes hands with everybody else exactly once. How many handshakes are there?

SOLUTION #1

Let us use our **visual-representation** strategy, by drawing a diagram. The 10 points, (no three of which are collinear), represent the 10 people. Begin with the person represented by point A.

We join A to each of the other nine points, indicating the first 9 handshakes that take place.

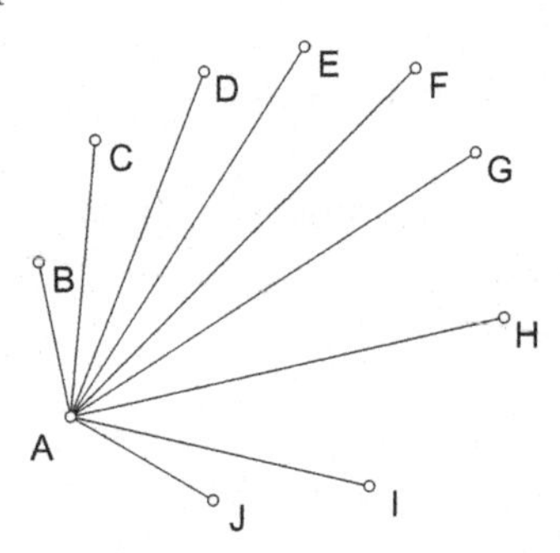

Now, from B there are 8 additional handshakes (since A has already shaken hands with B, and AB is already drawn). Similarly, from C there will be 7 lines drawn to the other points (AC and BC already drawn), from D there will be 6 additional lines or handshakes, and so on. When we reach point I, there is only one remaining handshake to be made, namely, I with J, since I has already shaken hands with A, B, C, D, E, F, G, and H. Thus, the sum of the handshakes equals $9+8+7+6+5+4+3+2+1=45$. In general, this is the same as using the formula for the sum of the first n natural numbers, $\frac{n(n-1)}{2}$, where $n \geq 2$. (Notice that the final drawing will be a decagon with all of its diagonals drawn.)

SOLUTION #2

We can approach the problem by **accounting for all the possibilities.** Consider the grid shown, which indicates persons $A, B, C, \ldots, H, I, J$, shaking hands with one another. The diagonal with the X's indicates that people cannot shake hands with themselves.

	A	B	C	D	E	F	G	H	I	J
A	X									
B		X								
C			X							
D				X						
E					X					
F						X				
G							X			
H								X		
I									X	
J										X

The remaining cells indicate doubly all the other handshakes (i.e., A shakes hands with B and B shakes hands with A). Thus, we take the total number of cells, (10^2) minus those on the diagonal (10), and divide the result by 2. In this case, we have $\frac{100-10}{2} = 45$.

In a general case for the $n \times n$ grid, the number would be $\frac{n^2-n}{2}$, which is equivalent to the formula $\frac{n(n-1)}{2}$, shown above.

SOLUTION #3

Let us now examine the problem by **adopting a different point of view.** Consider the room with 10 people, each of whom will shake nine other people's hands. This seems to indicate that there are 10×9 or 90 handshakes. But we must divide by 2 to eliminate the duplication, (since, when A shakes hands with B, we may also consider that as B shaking hands with A) and $\frac{90}{2} = 45$.

SOLUTION #4

Let us try to solve the problem by **looking for a pattern**. In the table shown below, we list the number of handshakes occurring in a room as the number of people increases.

Number of People in a Room	Number of Handshakes for Additional Person	Total Number of Handshakes in Room
1	0	0
2	1	1
3	2	3
4	3	6
5	4	10
6	5	15
7	6	21
8	7	28
9	8	36
10	9	45

The third column, which is the total number of handshakes, gives a sequence of numbers known as the *triangular numbers*, whose successive differences increase by 1 each time. It is, therefore, possible to simply continue the table until we reach the corresponding sum for the 10 people. We note that the pattern at each entry is one-half the product of the number of people on that line and the number of people on the previous line.

SOLUTION #5

We can approach the problem by a careful use of the **organizing data** strategy. The chart shown below shows each of the people in the room and the number of hands they have to shake each time, given that they have already shaken the hands of their predecessors, and do not shake their own hands. Thus, person number 10 shakes 9 hands, person number 9 shakes 8 hands, and so on. Finally, we reach person number 2, who only has one person's hand left to shake, and person number 1 has no hands to shake because everyone already shook his hand. Again the sum is 45.

Organized Data										
Person Number	10	9	8	7	6	5	4	3	2	1
Number of Handshakes	9	8	7	6	5	4	3	2	1	0

SOLUTION #6

We may also combine **solving a simpler problem** with **visual representation (drawing a picture)**, **organizing the data**, and **looking for a pattern**. Begin by considering a figure with one person, represented by a single point. Obviously, there will be 0 handshakes. Now, expand the number of people to 2, represented by 2 dots. There will be 1 handshake. Again, let us expand the number of people to 3. Now, there will be 3 handshakes needed. Continue with 4 people, 5 people, and so on.

Number of People	Number of Handshakes	Visual Representation
1	0	A
2	1	 A B
3	3	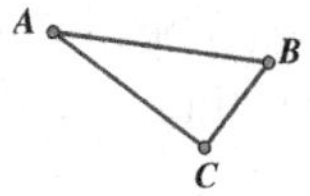 A B C
4	6	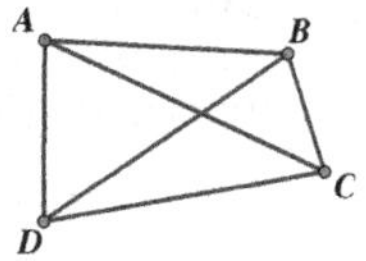 A B C D
5	10	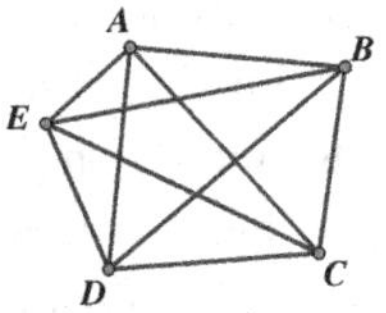 A B E C D

The problem has now become a geometry problem, where the answer is the number of sides and diagonals of an "n-gon". Thus, for 10 people we have a decagon, and the number of sides, $n = 10$. For the number of diagonals, we may use the formula

$$d = \frac{n(n-3)}{2}, \qquad \text{where } n > 3.$$

$$d = \frac{(10)(7)}{2} = 35.$$

Thus, the number of handshakes $= 10 + 35 = 45$.

SOLUTION #7

Of course, some readers might simply recognize that this problem could easily be resolved by applying the combinations formula of 10 things taken 2 at a time.

$${}_{10}C_2 = \frac{10 \cdot 9}{1 \cdot 2} = 45.$$

This solution, however, while quite efficient, brief, and correct, hardly utilizes any mathematical thought (other than application of a formula), and avoids the entire problem-solving approach. Although it is a solution that should be discussed, we find that the other solutions allowed us to demonstrate a variety of strategies — which is why we used this particular problem.

We suggest that you read the book through, work out the problems, and become familiar with all of the strategies. In this way you can develop your own set of problem-solving strategies, which become the basic tools of the problem-solving process. To those of you for whom problem solving is new, we hope the problems in the book will arouse your interest, and encourage you to delve further into this most interesting and necessary aspect of mathematics. To those who have been interested in critical thinking and problem solving for some time, we hope that you will find some new, interesting and unusual problems to intrigue you. Above all, enjoy!

Chapter 1

Logical Reasoning

It must seem rather redundant to have a chapter entirely devoted to a strategy referred to as logical reasoning. Regardless of the strategy being used to solve a problem, it would seem that some logical thinking must permeate the use of all the strategies. After all, to many people, problem solving is almost synonymous with logical reasoning, or logical thought. So, why have this chapter and why isolate this strategy?

In everyday life, we rely on logical reasoning when arguing a point with someone. After all, when we are having any kind of debate, we expect certain arguments to generate specific responses. At work, you can use a logical chain of arguments to change the way something is done in the office. We use logical reasoning to generate a chain of statements that, we hope, will lead to the conclusion we desire. In a courtroom, for instance, lawyers use logical reasoning to make their case for a desired verdict. If we are to meet someone in two days and today is Saturday, logic tells us we are meeting him, or her, on Monday.

In problem solving in mathematics, there are some problems that essentially do not involve any of the other strategies we normally use — some of which are presented in this book. Instead, they require us to arrive at a conclusion, which is reached by careful thought, and making a series of statements that follow one another in a logical chain of reasoning. For example, let us look at the following problem.

Find all pairs of prime numbers whose sum is 741.

Many people would make a list of all of the prime numbers less than 741 and search for those pairs that add up to 741. However, we can simplify our work with some logical reasoning. If the sum of two numbers is an odd number, one of the addends must be odd, the other must be even. But there is only one even prime number, namely 2. Therefore, the other number must be 739 (and 739 is a prime number). We have found all pairs that meet the given requirements.

Let us consider another problem where logical reasoning enables us to solve it.

> A palindromic number is one that reads the same forwards or backwards. Some examples of 3-digit and 4-digit palindromes are 373, and 8668. Maria wrote all the 3-digit palindromes on slips of paper and put them into a large box. Miguel wrote all the 4-digit palindromes on slips of paper and put them into the same box. The teacher stirred them all up, mixed them well, and asked Laura to pull one slip from the box without looking. What is the probability that she chose a 4-digit palindrome?

One method would be to write out all the 3-digit and 4-digit palindromes, count them all, and figure out the requested probability. This would work, even if it were somewhat time consuming. But if we use our logical reasoning strategy, we can simplify our work as follows. One example of a 3-digit palindrome might be 373. To make it a 4-digit palindrome all we have to do is repeat the middle digit, to obtain 3773. In fact, we can make every 3-digit palindrome into a 4-digit palindrome by simply repeating the middle digit once. Thus, the number of 4-digit palindromes is the same as the number of 3-digit palindromes, and so the probability of picking a 4-digit palindrome is one out of two, or $\frac{1}{2}$.

We will consider another example of how simple logical reasoning makes the solution of a problem rather simple.

> On a shelf in the florist's store, there are three boxes of ornamental bows to put on gift wrapped boxes. Mark went to put the three labels — "Red", "White", and "Mixed" (red and white) on the boxes. Unfortunately, he put the labels back, but put all three on the wrong boxes. Because the boxes are on a high shelf, Mark cannot look into the boxes. He knows all three are mislabeled, and he wants to reach up and pick one bow from one of the boxes. From which box should he pick the one bow in order to label all three boxes correctly?

Let us do some logical reasoning here. First, notice that whatever we say about the box labeled "White" we can also say about the box labeled "Red". There is a kind of symmetry there. So, let Mark pick the single bow from the box marked "Mixed". If it is red, then he knows that this box is really the box containing only red bows, since it cannot be the "Mixed" box. Label it "Red". The box labeled "White" cannot be all white, so it must be the "Mixed" box. Finally, the box incorrectly labeled "Red" must be the white box.

Notice that each of these problems requires not much more than some logical reasoning and careful thinking to reach a solution. This is not to say that we do not require logical thinking when using the other problem-solving strategies; however, the problems presented in this chapter rely almost exclusively on logical reasoning to reach a solution effectively.

PROBLEM 1.1

Max begins counting the natural numbers forward as $1, 2, 3, 4, \ldots$, while Sam is counting at the same speed but in the opposite direction — counting backwards from the number x as follows $x, x-1, x-2, x-3, x-4, \ldots$. When Max says the number 52, Sam says the number 74. With which number (x) did Sam start with in his counting backwards procedure?

A Common Approach

Faced with this problem most people will probably try to simulate the situation being described; that is, carrying both accounting procedures simultaneously to see what would result. The difficulty is that since one would not know where to begin the backward counting, the forward counting — in a trial and error procedure — would most likely be employed. This would not only be confusing, but very difficult to carry out.

An Exemplary Solution

Here, we will employ some logical reasoning. As Max counts 52 numbers, Sam will also be counting 52 numbers. We can designate Sam's 52nd number as $x-51$. However, we know that this number is to be 74. Therefore, we can equate them as $x - 51 = 74$, and then $x = 125$.

PROBLEM 1.2

We have 100 kg of berries and water, where 99% of the weight is water. A while later the water content of the mixture is 98% water. How much do the berries weigh?

A Common Approach

A common wrong answer is that with an evaporation of 1% water, that 99% must be the berries, which would imply that the berries weigh 99 kg. This is wrong!

An Exemplary Solution

Here we will have to use some logical reasoning to ascertain what is required. Initially, the mixture is 99% water, meaning that it contains 99 kg of water and 1 kg of dry matter, or 1% of the berries' mass. The mass of the dry matter does not change: at the end of the drying process, its weight remains 1 kg. In the meantime, however, the proportion of the total mass that is not water has doubled, to 2%.

In order for something that has a fixed quantity (our 1 kg of dry material) to double in proportion (going from 1% to 2%), the total amount of stuff has to be cut in half. We began with 1% or $\frac{1}{100}$ dry, and ended with 2% or $\frac{2}{100}$ dry, which reduces to $\frac{1}{50}$ — meaning we end up with 1 kg of dry matter out of 50 kg total. Thus, we have 49 kg of water at the end.

PROBLEM 1.3

In a class experiment, Miguel rolls one ordinary 6-sided die repeatedly. Keeping track of each number he rolls, he decides to stop as soon as one number is rolled three times. Miguel stops after the 12th roll, and the sum of these rolls is 47. Which number occurred for the third time? (An ordinary 6-sided die has the numbers from 1 through 6 on its sides)

A Common Approach

One approach is to obtain a die and carry out the experiment. It will be difficult to get an exact sum of 47 in 12 rolls, but even if you did get the answer that would be the inelegant method!

An Exemplary Solution

Let us use our strategy of logical reasoning. After the 11th roll, no number had yet appeared three times, otherwise the experiment would have already ended. This means that five of the numbers appeared twice, and the other number once. Let us call this number M. If M were rolled on the

12th roll, the total would then have been $2(1+2+3+4+5+6) = 42$. Therefore, the total after 11 rolls is $42 - M$. If N is the number rolled for the third time, then $42 - M + N = 47$ and so $N - M = 5$. We know that N and M can only take on the numbers from 1 through 6. The only two numbers that allow a difference of 5 from these numbers are 6 and 1. Therefore, with this restriction the equation $N - M = 5$, has only one solution where $M = 1$ and $N = 6$. Thus, it is the number 6 which is rolled a third time.

PROBLEM 1.4

We are given a triangle, whose perimeter is numerically equal to its area. What is the radius of the inscribed circle of the triangle?

A Common Approach

This is a problem where the common solution would entail drawing the picture as shown in Figure 1.1, and trying various values to see which ones might approach a possible solution. Frustration would be expected with this approach.

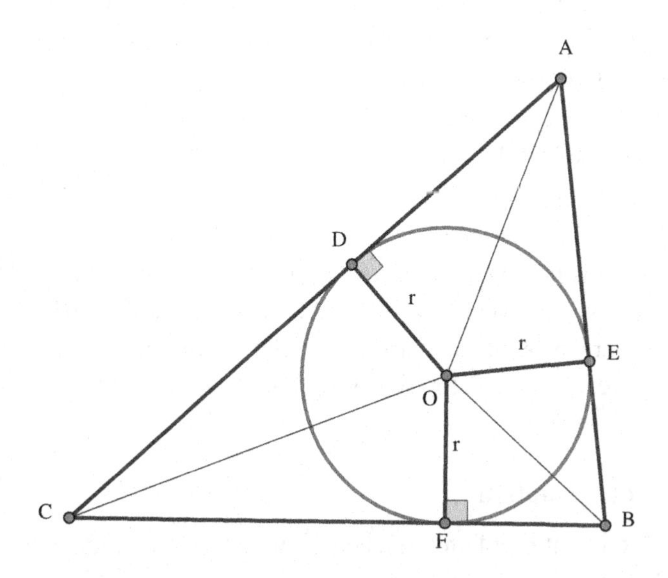

Figure 1.1

An Exemplary Solution

Here we would simply think logically and follow the dictates of the problem. We begin with triangle ABC whose perimeter would be $p = AB + BC + CA$. We will let point O be the center of the inscribed circle, with a radius of r. The area of triangle ABC is an equal to the sum of the areas of triangle AOB, triangle BOC and triangle COA, whose bases are: AB, BC, and CA respectively, and whose altitude is r. This gives us the following equation:

$$\begin{aligned}\text{Area } \Delta ABC &= \frac{1}{2}rAB + \frac{1}{2}rBC + \frac{1}{2}rCA \\ &= \frac{1}{2}r(AB + BC + CA) = \frac{1}{2}rp.\end{aligned}$$

Since we were told that the perimeter was numerically equal to the area, we get: $\frac{1}{2}rp = p$, so that $r = 2$.

PROBLEM 1.5

Presidents are elected every 4 years, in years that are multiples of 4. Some of these years are also perfect squares. How many presidential elections were there between 1788 and 2016 where the years are perfect squares? In what years did they occur?

A Common Approach

One way to approach this problem is to begin by taking all the 4-year election years that occurred between 1788 and 2016. Since 1788 is divisible by 4, it will be the first presidential election in the range under consideration. Thus we can list these years, (1788, 1792, 1796, ..., 2012, 2016) and then take the square root of each to see which of these years are actually perfect squares. A calculator will help but it is still a long and tedious process!

An Exemplary Solution

This is an excellent problem to make use of our logical-reasoning strategy. First of all, to be a multiple of 4, the years must be even, so we can disregard all the odd-numbered years. Furthermore, the square roots of these years must be in the 40's, since:

$$40^2 = 1600 \text{ (before the required range)}$$
$$42^2 = 1764 \text{ (before the required range)}$$
$$44^2 = 1936$$
$$46^2 = 2116 \text{ (after the required range).}$$

Only one year occurs *within* the given range and that is 1936. Therefore, there was only one presidential election year that was a perfect square, 1936.

PROBLEM 1.6

Jimmy flips two dimes at the same time. He keeps on doing this until at least one coin comes up heads (H). At this point, the game is over. What is the probability that both coins were heads at this last toss?

A Common Approach

The initial reaction is to actually flip two dimes and see what the results would be, after a large number of tosses. However, as in most probability experiments, the sample space is often too small to predict the result with any degree of accuracy.

An Exemplary Solution

Let us make use of our logical-reasoning strategy. In performing this experiment, the previous tosses of the coins are all totally irrelevant. Only one case where a head (H) appears is important. So, we can simply examine this last toss of the coins. The four possibilities are:

HH HT TH TT.

Of these tosses, three have at least one head. Only one has no heads; we can ignore this one. The only one with both H is HH. Thus, the probability is $\frac{1}{3}$.

PROBLEM 1.7

Some breeds of pigs are born with tails that have two curls. Others are born with tails that have 3 curls. A farmer sends his children to count how many pigs are in the pig pen. Since they are mathematically inclined, the children report back that the number of 2-curled pigs and the number of 3-curled pigs are each a prime number, and the total number of curls is 40. How many pigs are there in the farmer's pigpen?

A Common Approach

If we let x equal the number of pigs having 2-curled tails, and y equal the number of pigs having 3-curled tails, we can set up the equation $2x + 3y = 40$. This is a single equation with two variables. The numbers are small enough so that we can substitute different values for x and y until we find a set that satisfies. However, since we know that x and y are both prime numbers, it limits the choices we must use to the following: 19, 17, 13, 11, 7, 5, 3, and 2. Nevertheless, the process is quite long, tedious, and cumbersome.

An Exemplary Solution

If we let x equal the number of pigs having 2-curled tails, and y equal the number of pigs having 3-curled tails, then $2x + 3y = 40$, as we found before. However, let us use our logical-reasoning strategy and examine this equation. Since both 40 and $2x$ are even numbers, we know that y must also be even to get our sum (40) to be even. Since y is a prime number, it must be 2, (this is the only *even* prime number) and 3y must be 6. Therefore, we can now solve for x:

$$2x + 6 = 40,$$

$$2x = 34,$$

$$x = 17.$$

The farmer has $17 + 2$ or 19 pigs in the pen.

PROBLEM 1.8

We will refer to a number as "special" if it is divisible by the sum of its digits. Which of the following satisfy this condition?

11, 111, 1111, 11111, 111111, 1111111, 11111111, 111111111

A Common Approach

The usual approach is to find the sum of the digits of each number and then divide the number itself by that sum. For example, 11 must be divisible by $1 + 1$ or 2. It is not, so 11 is not a *special* number. Continue in the same manner for each of the other numbers. We have to solve eight small examples if we use this approach.

An Exemplary Solution

While the approach outlined above should eventually solve the problem, let us use our logical-reasoning strategy for a more elegant approach. First of all, it is apparent that all of the numbers shown are odd, since none ends in 2, 4, 6, 8, or 0. An even number of 1s would give an even sum. This eliminates those with and even number of ones: 11, 1111, 111111, and 11111111. Furthermore, 11111 is not divisible by 5, since it does not end in a 0 or a 5.

If we check 1111111 we find it is not divisible by 7. This leaves only two remaining numbers. We find that 111 is divisible by 3, the sum of its digits (i.e., $3 \cdot 37$). Also, 111111111 is divisible by 9 (i.e., $9 \cdot 12,345,679$). Thus, 111 and 111111111 are the only two "special" numbers in the original group.

PROBLEM 1.9

The smallest number that is divisible by the first nine counting numbers is 2,520. What would be the smallest number divisible by the first 13 counting numbers?

A Common Approach

The most common approach is to find all the factors of the first 13 counting numbers and multiply them. This would be time consuming, and the arithmetic is cumbersome. Remember, we must be careful not to repeat any factors (such as 8, since 4 and 2 would have already been used). However, this method would eventually get to the correct answer, if done carefully and correctly.

An Exemplary Solution

Let us make use of our logical-reasoning strategy. Obviously, all the factors of 1 through 9 (the first nine counting numbers) have been used when the product 2,520 was arrived at. We need, then, only consider the counting numbers 10, 11, 12 and 13 as we already have the number 2,520 covering the previous counting numbers. The factors of 10 ($5 \cdot 2$) and 12 ($4 \cdot 3$) have already been used. But, 11 and 13 are prime numbers, and have no factors other than themselves and 1. Thus, we multiply $2{,}520 \cdot 11 \cdot 13$ and find the smallest number that is divisible by the first 13 counting numbers to be 360,360.

PROBLEM 1.10

Al, Barbara, Carol and Dan each took a math test. Altogether, they answered 67 questions correctly, and every one had at least one question correct. Al had the most correct answers. Barbara and Carol together had 43 correct answers. How many answers did Dan have correct?

A Common Approach

A typical approach is to make a guess for each person. See if the statements made in the problem are followed and see if the guesses total 67 correct. This might lead to a correct response, but it would be largely dependent on clever guesses and not necessarily skill.

An Exemplary Solution

Let us use our logical-reasoning strategy. Since Barbara and Carol together had 43 correct, one of them must have had at least 22, while the other would have had 21 correct responses. Since Al had the most correct, and for a moment assuming the previous two assumptions about Barbara and Carol's responses, then Al had to have at least 23 correct responses. If we let Al have 23 correct responses, and Barbara with 22 and Carol with 21, together the 3 had $23 + 22 + 21 = 66$. This means Dan answered at most 1 question correctly. Since everyone had at least 1 correct, Dan must have had exactly 1 correct.

PROBLEM 1.11

Lisa was riding her bike, and was $\frac{3}{8}$ of the way across a bridge connecting points A and B, when she heard a train behind her approaching the bridge at 60 miles per hour. She did some quick mental calculations and found that she could *just* save herself by riding to either end of the bridge (to point A or point B) at her top speed. What is her top speed?

A Common Approach

Since the length of the bridge is not specified, let us assume a convenient (although a bit unrealistic) length of, say, 8 miles. If she rides back towards the beginning of the bridge (point A) at her speed of y miles per hour, she would travel 3 miles at a time of $\frac{3}{y}$ hours. During this time the train will be traveling a distance of x miles from point A in a time of $\frac{x}{60}$. This gives us the equation: $\frac{3}{y} = \frac{x}{60}$, or $xy = 180$.

If Lisa rides towards point B, in a similar manner we get the equation $\frac{5}{y} = \frac{x+8}{60}$, or $xy + 8y = 300$.

Combining these two equations, we get $8y = 300 - 180 = 120$, and then $y = 15$.

Lisa's top riding speed is, therefore, 15 miles per hour.

An Exemplary Solution

A more elegant solution is to use our logical reasoning strategy. If she can *just* reach either end of the bridge to save herself, we will let her ride away from the train towards point B. Now as the train arrives at point A, she will have ridden another $\frac{3}{8}$ of the way, for a total of $\frac{6}{8}$ of the bridge length (or $\frac{3}{4}$ of the length of the bridge). She can now ride the remaining $\frac{1}{4}$ of the bridge in the same time it takes the train to cover the entire bridge length. Thus, her speed is $\frac{1}{4}$ that of the train or 15 miles per hour.

PROBLEM 1.12

If $S = 1! + 2! + 3! + 4! + 5! + \cdots + 98! + 99!$, then what is the units' digit of the value of S?

Note: the symbol $n!$ represents $1 \cdot 2 \cdot 3 \cdot 4 \cdot \ldots \cdot (n-1) \cdot n$.

A Common Approach

Typically, when one is faced with a problem like this, the urge will be to work out each of the factorials and then add them to get the value of S. Needless to say, this is a tedious task, which would likely lead to arithmetic mistakes.

An Exemplary Solution

As we inspect the value of S and simplify it, we get the following:

$$S = 1! + 2! + 3! + 4! + 5! + \cdots + 98! + 99!$$
$$S = 1 + 2 + 2 \cdot 3 + 2 \cdot 3 \cdot 4 + 2 \cdot 3 \cdot 4 \cdot 5 + \cdots + 98! + 99!$$
$$S = 1 + 2 + 6 + 24 + 10k, \quad \text{where } k \text{ is a positive integer.}$$

We have reduced the terms from powers of 5! onward to $10k$, since 5! has a factor of 10 embedded. Any multiple of 5! will be a multiple of 10. Since 6! is a multiple of 5! and 7! is a multiple of 6!, for any n greater than 5, $n!$ will be a multiple of 10. Therefore, the units digit will be 0.

Chapter 2

Pattern Recognition

Some of the most inherent beauties of mathematics are the many patterns that occur in problems as we try to resolve them. The famous mathematician, W. W. Sawyer once said that mathematics could be thought of as a search for patterns. One of the major uses of mathematics is to predict things that occur in a regular manner. For example, how many scones will I need for 3 people? For 4 people? For 10 people? For n people?

Recognizing patterns is an important problem-solving skill. If you recognize a pattern when you look systematically at a set of specific examples, you can use that pattern to generalize what you have seen into a broader, more open solution to a problem. For example, when asked what are the next two numbers in the sequence 1, 2, 3, 6, 11, 20, 37, __, __, we have to examine the sequence to see if the numbers fit into any sort of pattern. After all, if the first three terms were 1, 2, 3, should not the next one be 4? But it is not!! Aha! We notice that each term after the 3rd is the sum of the three preceding numbers. (This is a Fibonacci-type sequence). That is $1 + 2 + 3 = 6$, $2 + 3 + 6 = 11$, $3 + 6 + 11 = 20$, and so on. If we continue in this manner, we find the next two numbers in the sequence to be $11 + 20 + 37 = \mathbf{68}$, and $20 + 37 + 68 = \mathbf{125}$.

Even young children make use of patterns. When youngsters start in school, they learn to count. They use patterns to help them count by 1s, then by 2s by 5s and so on. If we ask a second grader what number comes next in the sequence 3, 6, 9, 12, ... the child will ask himself, or herself, "What do I add to each number to get the next one?" This is an almost-natural use of a search-for-patterns strategy developed informally.

Most of us use patterns in our everyday lives. Some of these "patterns" involve mnemonics. Mnemonic is a word from the ancient Greeks *mneomnikos*, which means a memory device. Many of us are familiar with the musical mnemonic "**E**very **G**ood **B**oy **D**oes **F**ine", where the initial letters of each word form a pattern naming the notes that appear on the lines of a musical measure, namely E–G–B–D–F. We make use of patterns in remembering a code for a gym combination lock, a telephone number, or a license plate number. When we are looking to find a certain house number, we almost intuitively make use of the pattern that the odd numbers are usually on one side of the street, and the even numbers on the other side of the street — a very simple pattern, yet a valuable one.

The police make a wide use of patterns. If a series of crimes occur, the police look for an M.O. (*modus operandi)* of the criminals. That is, is there a pattern in the process of these crimes?

A doctor usually depends on patterns of behavior to determine an illness in a patient. After treating many cases of an illness, the physician can readily recognize the pattern of symptoms in diagnosing the illness in future cases.

The real power of the strategy of pattern recognition is best seen, when we use the strategy to resolve a problem situation, especially when it is not obvious that the strategy can be used to solve a particular problem. For instance, suppose we were asked to find the unit's digit of the number generated by 13^{23}. The most obvious approach is to use a calculator and raise 13 to the 23rd power. But this is a rather formidable task even if we have a calculator that can show the number of digit places for this extremely large number. Instead, we can examine the increasing powers of 13, and see if the unit's digits form any sort of a pattern that might help us to answer this question.

$$\begin{array}{ll}
13^1 = 1\underline{3} & 13^5 = 371{,}29\underline{3} \\
13^2 = 16\underline{9} & 13^6 = 4{,}826{,}80\underline{9} \\
13^3 = 2{,}19\underline{7} & 13^7 = 62{,}748{,}51\underline{7} \\
13^4 = 28{,}56\underline{1} & 13^8 = 725{,}731{,}72\underline{1}
\end{array}$$

Looks like the unit's digits for the powers of 13 form a sequence:

$$3, 9, 7, 1, 3, 9, 7, 1, \ldots$$

These occur in a cycle of 4. Thus, 13^{23} will have the same unit's digit as 13^3, or 7.

In fact, this problem raises an interesting question about patterns. Do *all* numbers have a cyclical pattern of the unit's digit for the powers of a particular number? Some we recognize right away. For example, powers of 5 always have a unit digit of 5 (That is, 5, 25, 125, 625, ...). This property of numbers is a most interesting one and quite valuable in solving problems by means of pattern recognition. You might try to determine the pattern of the unit's digit for powers of other numbers to see if a pattern evolves.

We must, however, add a word of caution. There are times, fortunately not often, where a pattern seems to be developing, but is not consistent throughout. One such pattern is where it would appear that every odd number beginning with the number 3 can be expressed as the sum of a power of 2 plus an odd number. When we try to verify this by example, we find that the "rule" seems to hold true up to the number 125. However, surprisingly, it does not hold true for the next odd number 127. Therefore, we must be careful when applying a pattern strategy in solving problems. This is clearly an exception, and should not distract from using this method of solving problems.

$$3 = 2^0 + 2$$
$$5 = 2^1 + 3$$
$$7 = 2^2 + 3$$
$$9 = 2^2 + 5$$
$$11 = 2^3 + 3$$
$$13 = 2^3 + 5$$
$$15 = 2^3 + 7$$
$$17 = 2^2 + 13$$
$$19 = 2^4 + 3$$

and so on

$$51 = 2^5 + 19$$

and so on

$$125 = 2^6 + 61$$
$$127 = ?$$
$$129 = 2^5 + 97$$
$$131 = 2^7 + 3.$$

We will now embark on a series of problems that will be more effectively solved when a pattern is recognized — especially where one is not necessarily expected.

PROBLEM 2.1

What is the unit's digit of the following expression $2^{2^{2^{2^{2^{2^{\cdot^{\cdot^{\cdot^{2}}}}}}}}}$ when there are 222 appearances of the digit 2 in the exponent?

A Common Approach

Unfortunately, there will be people who will feel that to evaluate this expression they have to work out all the values of this as it increases to the 222 powers. This cannot be a successful path to a solution!

An Exemplary Solution

Let us see if we can recognize a pattern as the powers of 2 increase according to the pattern described in the problem. As the powers of 2 increase, the unit's digits repeat in a pattern 2, 4, 8, 6. That is,

$$2^1 = 2$$
$$2^2 = 4$$
$$2^3 = 8$$
$$2^4 = 16$$
$$2^5 = 32$$
$$2^6 = 64$$
$$2^7 = 128$$
$$2^8 = 256$$

The exponent at the third step of our calculation below is a multiple of 4, and any exponent of a power of 2 that is a multiple of 4 will in turn produce a number, where the units digit is a 6.

$$2$$
$$2^2 = 4$$
$$2^{2^2} = 2^4 = 16$$
$$2^{2^{2^2}} = 2^{16} = 65{,}536$$
$$2^{2^{2^{2^2}}} = 2^{65{,}536} = 11579208923731619542357098500868790785326998\ 4665640564039457584007913129639936$$

Therefore, the unit's digit of our expression is 6.

PROBLEM 2.2

Each of the rectangular arrays that follow contains a certain number of dots. How many dots would be in Figure 49?

FIGURE 1 **FIGURE 2** **FIGURE 3** **FIGURE 4**

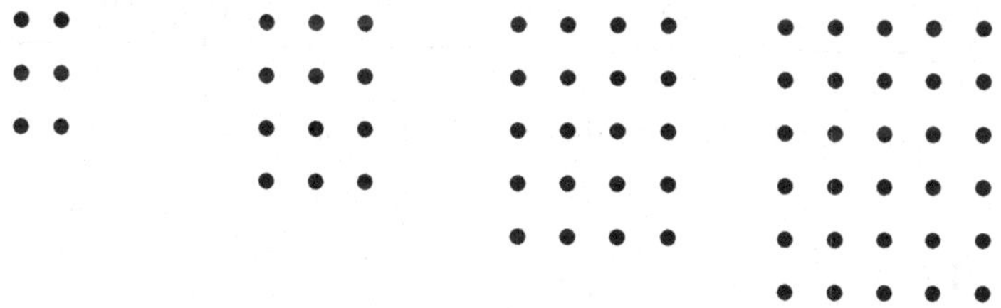

A Common Approach

The obvious approach is to continue drawing the dot arrays until we reach Figure 49. Then we can count the dots. This would take a long time and you would need a great deal of patience, not to mention a great deal of paper. Somehow, it seems like there must be a better way.

An Exemplary Solution

Let us try organizing the data and searching for a pattern. We will make a table of what we already know.

Figure Number	Height	Width	Total Number of Dots
1	3	2	6
2	4	3	12
3	5	4	20
4	6	5	30

Aha! There is a pattern. Each height is 2 more than the figure number and the width is 1 more than the figure number. So, for figure n,

Figure Number	Height	Width	Total Number of Dots
1	3	2	6
2	4	3	12
3	5	4	20
4	6	5	30
•	•	•	•
•	•	•	•
•	•	•	•
n	$(n+2)$	$(n+1)$	$(n+2)(n+1)$

Thus, for the 49th figure there will be $(51)(50) = 2{,}550$ dots.

PROBLEM 2.3

A circle can be cut into seven pieces with three straight lines. What is the maximum number of pieces that can be cut from a circle using seven straight lines?

A Common Approach

A typical approach to this problem is to take a circle and to draw seven straight lines through it, avoiding any three to become concurrent — that is, containing the same point. If one does it carefully, it should lead to a correct answer. However, making sure that you maximize the number of pieces could be a problem.

An Exemplary Solution

An interesting way to approach this problem would be to see if there is a pattern that might appear as we increase the number of lines cutting the circle into pieces — keeping in mind that no three lines could contain the same point. It is clear that, if we draw one line through a circle, there will be two parts into which the circle is partitioned. If we draw two lines, the circle will be divided into four pieces. The table below will show you the number of pieces into which a circle can be cut by a given number of lines, no three of which contain the same point.

Number of Lines	Number of Pieces	Differences
1	2	
		2
2	4	
		3
3	7	
		4
4	11	

A pattern seems to be developing amongst the differences — where they seem to be increasing by one each time. Therefore, testing the next case, where five lines seem to be creating 16 pieces, we can, perhaps, conclude the following table applying the pattern.

Number of Lines	Number of Pieces	Differences
1	2	
		2
2	4	
		3
3	7	
		4
4	11	
		5

(*Continued*)

(*Continued*)

Number of Lines	Number of Pieces	Differences
5	16	
		6
6	22	
		7
7	29	

Therefore, a circle cut by seven chords can produce 29 pieces.

PROBLEM 2.4

We are given a map showing the direction of movement along the streets as indicated by the arrows in Figure 2.1.

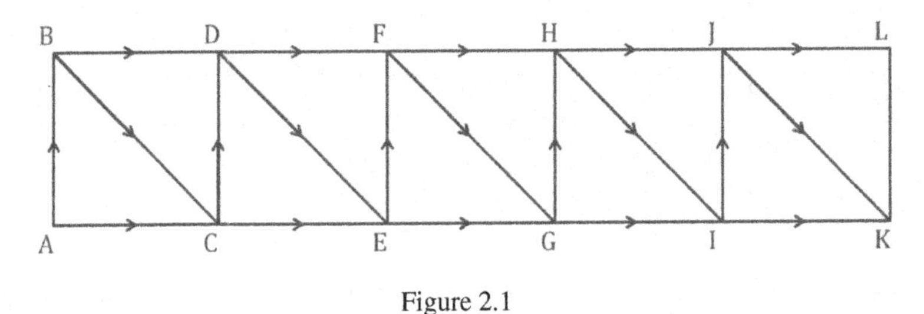

Figure 2.1

How many different routes are there to get from point A to point L?

A Common Approach

The most common approach is to simply count the various possible routes. That is, to consider the different routes one at a time, and then count the results. For example, one route is A–B–C–D–E–F–G–H–I–J–K–L. Another is A–C–D–E–G–K–L and so on. However, as you can readily see, this can be quite cumbersome and might be difficult to count and avoid any duplication of all the different routes. There are quite a few!

An Exemplary Solution

Let us use our strategy of finding a pattern. Suppose we wish to go from A to B. There is only one way (A–B). To go from A to C yields 2 ways (A–B–C, or A–C). To go from A to D yields 3 ways, namely,

(A–B–D, A–C–D, A–B–C–D). If we continue in this manner, we get the following number of ways for each location through point F.

To Reach Point	Number of Paths
A	1
B	1
C	2
D	3
E	5
F	8

We show this in Figure 2.2.

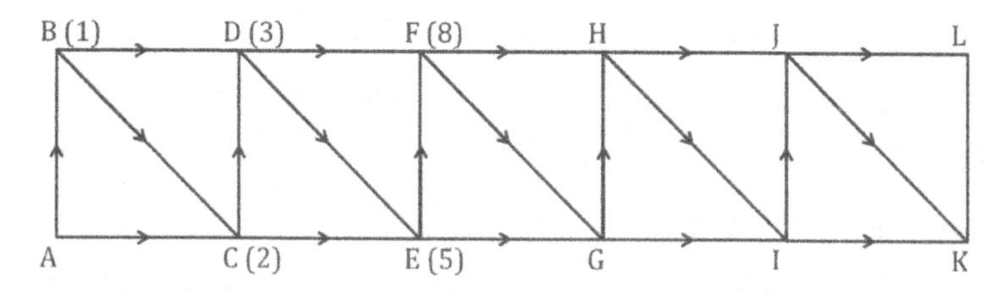

Figure 2.2

The sequence of numbers 1, 2, 3, 5, 8, 13 forms a Fibonacci-type sequence, which is a famous sequence of numbers first popularized in the Western world by Leonardo of Pisa (also known as Fibonacci) in the year 1202. The sequence begins with 1 and 1, and then continues to generate subsequent numbers by taking the sum of the previous two numbers. When we continue this sequence to the point L, we get the following:

1, 2, 3, 5, 8, 13, 21, 34, 55, 89, 144.

Therefore, we have used this pattern to find that there are 144 different ways to go from point A to point L.

PROBLEM 2.5

Johnny takes a sheet of paper from his notebook and tears it in half. He puts the two pieces of paper together, one on top of the other, and tears them in half again. He places the resulting pieces of paper on top of one another and tears them in half again. If Johnny could repeat this process for a total of 20 times, how thick would the pile of papers be? (Assume the original sheet of paper was 0.001 inches thick)

A Common Approach

We can make a table and simulate the action.

Number of sheets in the pile	Tear #	Total in pile	Thickness
1	1	2	0.002
2	2	4	0.004
4	3	8	0.016
8	4	16	0.032

and so on. Eventually, we can complete the table for a total of 20 tears and find the answer.

An Exemplary Solution

Let us use our strategy of finding a pattern to solve the problem. After 1 tear, the pile is 2 layers thick After 2 tears, it is 4 layers thick. After 3 tears it is 8 layers thick. In exponential form the thickness can be written as $2^1, 2^2, 2^3, \ldots,$ which has 2^n as a general term. After 20 tears, the thickness will be $0.001 \cdot 2^{20}$ or about 1049 inches which is about 87 feet; that is why the problem said "*If* Johnny could repeat the process 20 times".

PROBLEM 2.6

How many squares of all sizes are there on a standard 8-square by 8-square checkerboard?

A Common Approach

The immediate reaction is to say 8×8 or 64 squares. However, the phrase "of all sizes" suggests there may be others as well. The mathematical approach is to try to count how many square regions there are of all sizes, located on the 64-square checkerboard, namely, $2 \times 2, 3 \times 3, 4 \times 4$, etc. This is awkward and quite difficult to visualize, since many squares overlap. Furthermore the counting process could become confusing, thus making it a rather tedious and troublesome method of solution.

An Exemplary Solution

We can use the strategy of looking for a pattern together with a table to organize the data. If we start with a 1-square by 1-square checkerboard, there is obviously only one square, that is the 1×1 square. On a 2-square by 2-square checkerboard we see that there are 4 of the 1-square by 1-square, and 1 of the 2 by 2 squares for a total of 5 squares. Organizing the data in a table as we increase the size of our checkerboard from 1×1 to 2×2 to 3×3 and so on, we obtain the table below.

Board size	1×1	2×2	3×3	4×4	5×5	6×6	7×7	8×8	Total
1×1	1	—	—	—	—	—	—	—	1
2×2	4	1	—	—	—	—	—	—	5
3×3	9	4	1	—	—	—	—	—	14
4×4	16	9	4	1	—	—	—	—	30
5×5	25	16	9	4	1	—	—	—	55
6×6	36	25	16	9	4	1	—	—	91
7×7	49	36	25	16	9	4	1	—	140
8×8	64	49	36	25	16	9	4	1	204

As we continue the table till we get to an 8 by 8 checkerboard, we notice the pattern of squares moving along in each row so that we can then conclude that there are 204 squares of all sizes on an 8×8 square checkerboard.

Once we have constructed the table shown above, we notice the amazing number of patterns. The perfect squares abound throughout the table. If we examine the "Total" column and take the differences between consecutive terms, we get another interesting sequence:

$$5 - 1 = 4$$

$$14 - 5 = 9$$

$$30 - 14 = 16$$

$$55 - 30 = 25$$

$$91 - 55 = 36$$

$$140 - 91 = 49$$

$$204 - 140 = 64$$

Again, the perfect squares appear. If we now take the second differences — that is the difference between the squares — we obtain the sequence of odd numbers beginning with 5.

$$9 - 4 = 5$$
$$16 - 9 = 7$$
$$25 - 16 = 9$$
$$36 - 25 = 11$$
$$49 - 36 = 13$$
$$64 - 49 = 15.$$

Patterns are not only very useful in solving problems, as we noted with the original problem here, but they are also an aspect of mathematics that adds to its beauty.

PROBLEM 2.7

The table shown below continues indefinitely. What will be the middle letter in the 30th row?

ROW 1	G L A D Y S G
ROW 2	L A D Y S G L
ROW 3	A D Y S G L A
ROW 4	D Y S G L A D.

A Common Approach

We can continue writing out the letters in each row until we reach row 30. Then we can find the middle letter. It is a rather clumsy method, but it will give the correct answer.

An Exemplary Solution

This would be a classic example of how seeking a pattern leads efficiently to a solution. Let us continue the pattern for 4 more rows

ROW 1	G L A D Y S G
ROW 2	L A D Y S G L
ROW 3	A D Y S G L A
ROW 4	D Y S G L A D
ROW 5	Y S G L A D Y
ROW 6	S G L A D Y S
ROW 7	G L A D Y S G
ROW 8	L A D Y S G L ...

Since there are 6 letters in the set, the rows will repeat after every 6 letters. Furthermore, since 30 is an exact multiple of 6, the middle letter in row 30 is the same as the middle letter in row 6, or A. The use of the recognizing a pattern strategy resolves the problem quite easily.

PROBLEM 2.8

Find the unit's digit of the number equal to each of the following:

(a) 8^{19},

(b) 7^{197}.

(Naturally, this should be done without a calculator or a computer.)

A Common Approach

Some people may begin to approach this problem by entering the powers of 8 into their calculators. However, they should soon realize that most calculators would not permit them to arrive at an answer of that magnitude, since the display will run out of digit space before they reach their goal.

An Exemplary Solution

We must look for another approach. Let us examine the increasing powers of 8 and see if there is a pattern in the last digits, which may be of use to us.

$8^1 = \underline{\mathbf{8}}$	$8^5 = 32{,}76\underline{\mathbf{8}}$	$8^9 = 134{,}217{,}72\underline{\mathbf{8}}$
$8^2 = 6\underline{\mathbf{4}}$	$8^6 = 262{,}14\underline{\mathbf{4}}$	$8^{10} = 1{,}073{,}741{,}82\underline{\mathbf{4}}$
$8^3 = 51\underline{\mathbf{2}}$	$8^7 = 2{,}097{,}15\underline{\mathbf{2}}$	$8^{11} = 8{,}589{,}934{,}59\underline{\mathbf{2}}$
$8^4 = 4{,}09\underline{\mathbf{6}}$	$8^8 = 16{,}777{,}21\underline{\mathbf{6}}$	$8^{12} = 68{,}719{,}476{,}73\underline{\mathbf{6}}$.

Notice the pattern that emerges — the unit's digits repeat in cycles of four powers. It would appear that we can apply this pattern to our original problem. The exponent we are interested in is 19, which gives a remainder of 3 when divided by 4. Thus, the terminal digit of 8^{19} should be the same as those of 8^{15}, 8^{11}, 8^{7}, 8^{3}, which we recognize as 2.

By the way, for the skeptical reader, here is the actual value of $8^{19} = 144,115,188,075,855,87\underline{2}$.

In a similar way, we can now examine the increasing powers of 7 and see if there is a pattern, which may be of use.

$$
\begin{array}{lll}
7^1 = \underline{\mathbf{7}} & 7^5 = 16{,}80\underline{\mathbf{7}} & 7^9 = 40{,}353{,}60\underline{\mathbf{7}} \\
7^2 = 4\underline{\mathbf{9}} & 7^6 = 117{,}64\underline{\mathbf{9}} & 7^{10} = 282{,}475{,}24\underline{\mathbf{9}} \\
7^3 = 34\underline{\mathbf{3}} & 7^7 = 823{,}54\underline{\mathbf{3}} & 7^{11} = 1{,}977{,}326{,}74\underline{\mathbf{3}} \\
7^4 = 2{,}40\underline{\mathbf{1}} & 7^8 = 5{,}764{,}80\underline{\mathbf{1}} & 7^{12} = 13{,}841{,}287{,}20\underline{\mathbf{1}}.
\end{array}
$$

Following this pattern we have, the exponent $\frac{197}{4}$ leaves a remainder of 1 and so the unit's digit of 7^{197} should be the same as 7^1 which is 7. We can check this answer, if we have time, and we will find that

$$
\begin{aligned}
7^{197} = {} & 305009862720535194606965003259965412822716867351901 85 \\
& 597522274297478500779662572162607529489531673616014 \\
& 767487616753102548289155520943414542713569292535908 \\
& 264249143207.
\end{aligned}
$$

PROBLEM 2.9

To make a 1×1 square, requires 4 toothpick as shown in Figure 2.3.

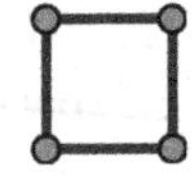

Figure 2.3

To make a 2×2 square requires 12 toothpicks (Figure 2.4)

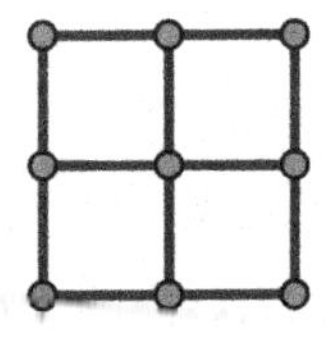

Figure 2.4

How many toothpicks will be needed to make a 7×7 square?

A Common Approach

We can actually draw the 7 × 7 square, and then count the number of toothpicks needed. This procedure will work, but it is cumbersome and requires a carefully made drawing.

An Exemplary Solution

We will begin by drawing a few of the smaller squares to see if there might be a pattern evolving. Draw the 3 × 3 and the 4 × 4 squares (see Figures 2.5 and 2.6), and let us see if there might be a pattern to help solve the problem.

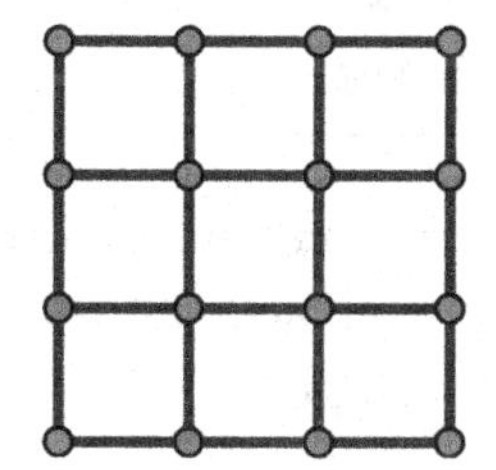

Figure 2.5

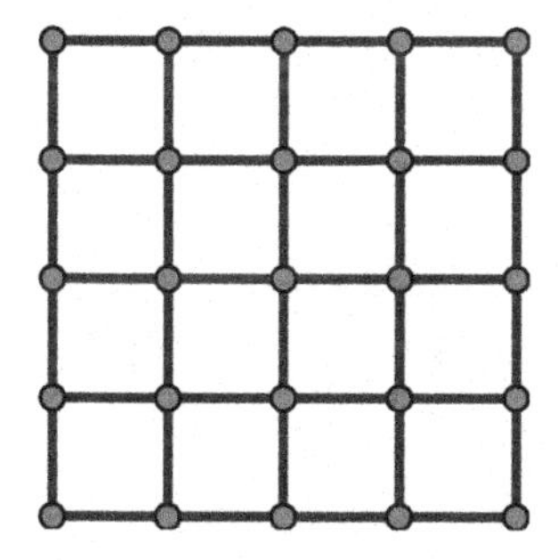

Figure 2.6

Let us see what we know.

Square	Number of toothpicks	How many more
1 × 1	4	—
2 × 2	12	8
3 × 3	24	12
4 × 4	40	16

Aha! As the size of the large square increases by 1, the number of toothpicks needed increases by 4. We can continue the table:

Square	Number of toothpicks	How many more
1×1	4	—
2×2	12	8
3×3	24	12
4×4	40	16
5×5	60	20
6×6	84	24
7×7	**112**	28

The table shows that the third column indicates an increase by 4's. We can compute the number of toothpicks by working backwards from that column. You would use 112 toothpicks.

Chapter 3

Working Backwards

The very name of this strategy sounds confusing to most people. It is a very unnatural way of doing things. When most of us went to school, we were taught to solve mathematical problems in a direct, straightforward manner. And yet, working backwards is the way many real-life problems are often resolved. As a simple example, suppose you had to pick up your child from football practice at exactly 5:00 p.m. At what time should you leave? Well, let us say it take 30 minutes to get to the ballpark. We would better leave a 5-minute safety valve. Okay, then we need to leave 35 minutes earlier, or not later than 4:25 p.m. Without even thinking about it, we were working backwards! Of course, this is a very simple example of this strategy.

To further make us aware of this sort of thinking, we will consider another example. Suppose there were a car accident. The police would work backwards from the scene of the accident. Who hit whom? Which car swerved? How far were the skid marks? Who had the right-of-way? This is just another example of working backwards.

When using our working backwards strategy, we usually start with the end of the problem, or the "answer." From that point, we reverse the operations called for. Thus, if the problem says "increase by 2," we will "decrease by 2," or subtract 2. After all, if we increased something by 2 we should diminish it by the same 2 to get back to the previous step. Similarly, if multiplying by 3 is mentioned, then, when working backwards, we would divide by 3. Let us look at a typical problem.

> Maria's average for 11 tests is 80. When the teacher computes her final average, she is very generous; she eliminates the lowest score. In this case, she dropped Maria's test score of 30. What is Maria's final average?

We will work backwards from her average. An average (or arithmetic mean) is usually computed by adding all the scores and dividing by the number of scores. If her average for 11 tests was 80, then the total for these 11 tests must be 11×80 or 880. (Notice that we multiplied by 11 — reversing the original operation of dividing by 11.) Subtracting the test score of 30 that the teacher removed and subtracting one test score, we find the total for her remaining 10 tests to be 850. Her final average is

$$850 \div 10 \quad \text{or} \quad 85.$$

Let us try another problem that can benefit from our working backwards strategy.

> David just came back from playing four games of baseball cards. He now has 45 cards in his package of cards. When I asked how he did, he told me he had lost one-half of his cards in the first game. In the second game, he won 12 times what he had at that time. In the third game he won 9 cards. The fourth game was a tie so no cards changed hands. How many cards did he start with?

We can construct a series of equations following the action of the problem. But, let us see how our working-backwards strategy comes into play. We are given the final result (45 cards) and asked to find the start. This is the "trademark" of a typical problem that lends itself nicely to the working-backwards strategy. He finished with 45 cards. The fourth game was a tie, so he still had 45 cards at the end of the third game. In the third game, he won 9 cards, so he must have had 36 cards at the end of the second game. In the second game, he won 12 times what he had, so he must have had 3 cards at the end of the first game. In the first game, he lost one-half of his cards, so he must have started with 6 cards. The working-backwards strategy allowed us to solve this problem rather easily.

PROBLEM 3.1

The sum of two whole numbers is 2. The product of these same two numbers is 5. Find the sum of the reciprocals of these two numbers.

A Common Approach

The problem immediately suggests forming two equations in two variables:

$$x + y = 2,$$
$$xy = 5.$$

These two equations can be solved simultaneously by using the quadratic formula, which is $x = \frac{-b\pm\sqrt{b^2-4ac}}{2a}$, for $ax^2 + bx + c = 0$. However, the method yields complex values for both x and y, namely, $1 + 2i$, and $1 - 2i$. Following the requirements of the original problem, we now need to take the sum of the reciprocals of these two roots.

$$\frac{1}{1+2i} + \frac{1}{1-2i} = \frac{(1-2i)+(1+2i)}{(1+2i)(1-2i)} = \frac{2}{5}.$$

We should emphasize here that there is nothing wrong with this method, it is just not the most elegant way to solve this problem.

An Exemplary Solution

Before embarking on a problem it usually makes sense to step back from it and see what is being required. Curiously, this problem is not asking for the values of x and y, but rather the sum of the reciprocals of these two numbers. That is, we seek to find $\frac{1}{x} + \frac{1}{y}$. Using a strategy of working backwards, we could ask ourselves where does this lead us. Adding these two fractions could give us this answer. Therefore, $\frac{1}{x} + \frac{1}{y} = \frac{x+y}{xy}$. At this point the required answer is immediately available to us since we know the sum of the numbers is 2, and the product of the numbers is 5. We merely substitute these values in the last fraction to get $\frac{1}{x} + \frac{1}{y} = \frac{x+y}{xy} = \frac{2}{5}$, and our problem is solved.

PROBLEM 3.2

Lauren has an 11-liter can and a 5-liter can. How can she measure out exactly 7 liters of water?

A Common Approach

Most people will simply guess at the answer, and keep "pouring" back and forth in an attempt to arrive at the correct answer, a sort of "unintelligent" guessing and testing.

An Exemplary Solution

However, the problem can be solved in a more organized manner by using the strategy of working backwards. We need to end up with 7 liters in the 11-liter can, leaving a total of 4 empty liters in the can. But, where do 4 empty liters come from? (Figure 3.1)

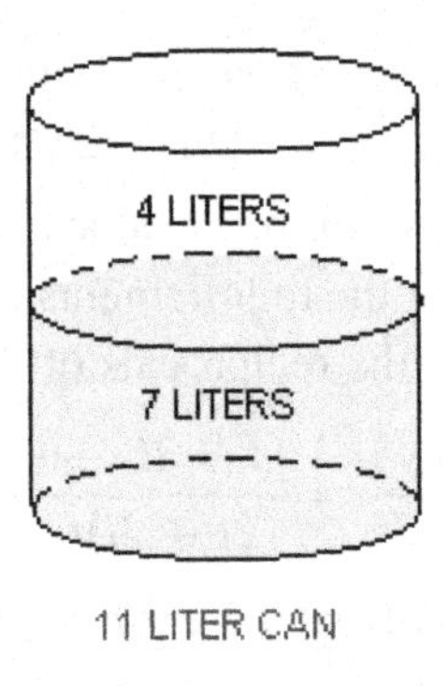

Figure 3.1

To obtain 4 liters, we must leave 1 liter in the 5-liter can. Now, how can we obtain 1 liter in the 5-liter can? Fill the 11-liter can and pour from it twice into the 5-liter can, discarding the water. This leaves 1-liter in the 11-liter can. Pour the 1-liter into the 5-liter can (Figure 3.2).

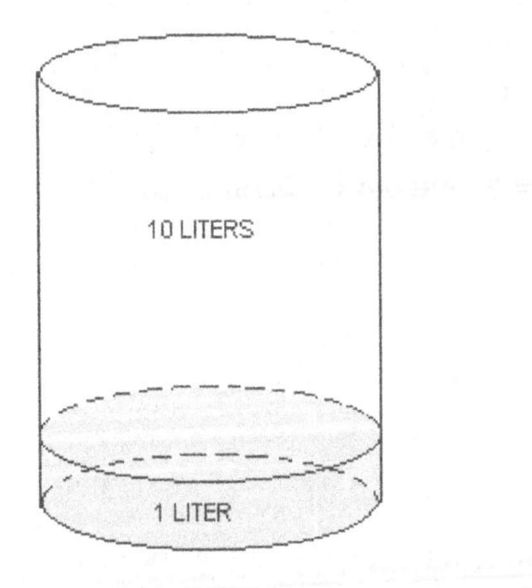

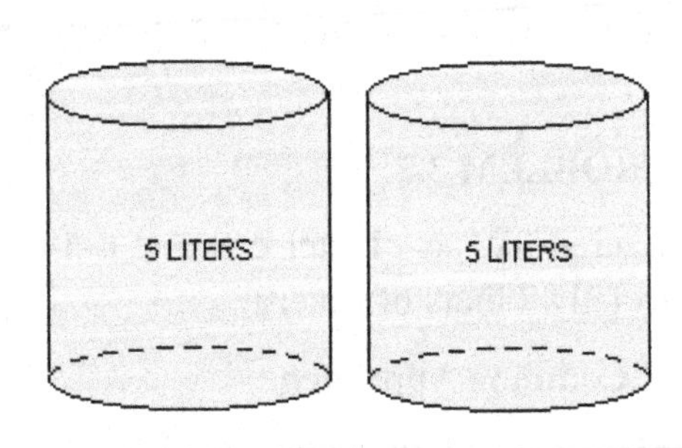

Figure 3.2

Now, fill the 11-liter can and pour off the 4 liters needed to fill the 5-liter can. This leaves the required 7 liters in the 11-liter can (Figure 3.3).

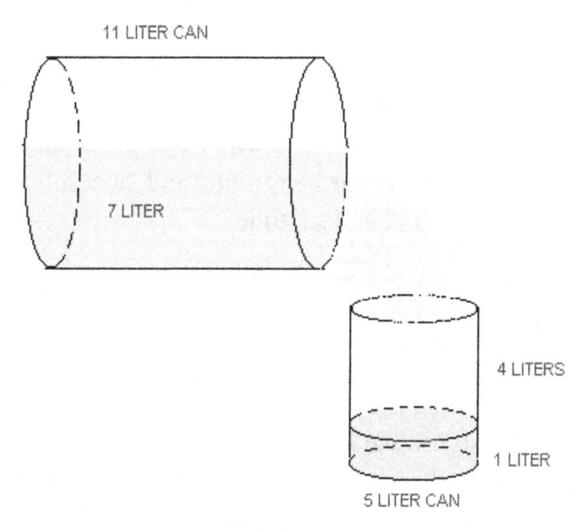

Figure 3.3

Note that problems of this sort do not always have a solution. That is, if you wish to construct additional problems of this sort, you must bear in mind that a solution only exists when the difference of multiples of the capacities of the two given cans can be made equal to the desired quantity. In this problem, $2(11) - 3(5) = 7$.

This concept can lead to a discussion of parity. We know that the sum of two like parities will always be even (that is, even + even = even; odd + odd = even) whereas the sum of two unlike parities will always be odd (odd + even = odd). Thus, if two even quantities are given, they can never yield an odd quantity.

PROBLEM 3.3

The definition of a palindrome is a number that reads the same forwards and backwards. For example, the numbers 66, 595, 2332, 7007 are all palindromes. Jack's teacher assigned the class to find the sum of the first 15 counting numbers. Jack used his calculator and added the numbers from 1 through 15. He was a bit surprised when his answer turned out to be a palindrome. Jack did not realize that he had omitted one of the numbers. Which number did Jack forget to enter?

A Common Approach

The usual approach is to try all possible combinations of additions, leaving out one number each time until the sum of the 14 numbers chosen yields a palindrome. This brute force method will work, especially if you use a calculator. However, it is very time consuming and, hopefully, you will not leave out more than one number each time.

An Exemplary Solution

Let us try a different approach, by first examining what the sum of the first 15 counting numbers should be. Although we can use a well-known formula for finding the sum of an arithmetic progression, namely, $S = \frac{(n)(n+1)}{2}$, we could also use the extremely clever way that the young Carl Friedrich Gauss found the sum of an arithmetic progression at age 10. Rather than to add the numbers in the given order:

$1 + 2 + 3 + \cdots + 14 + 15$, he simply added the first number to the last number, then the second number to the next-to-the-last number and so on to find that he had seven pairs of 16 and the 8 in the middle, giving a sum of $7 \cdot 16 + 8 = 120$.

Since he left out one number and got a palindrome that must have been 111. Now you may wonder, why could he have not gotten another palindrome, such as 101? In order for him to have gotten 101, having left out one number in his addition, that number would have been 19, which was not on his list of numbers to be added, that is 1–15. Therefore, the number that he omitted in his addition must have been the number 9.

PROBLEM 3.4

Mrs. Suttner baked some cookies for Bertha's lunches. The first day, Bertha ate one-half of the cookies. On the second day, she ate one-half of what was left. On the third day, she ate one-fourth of what was left. On the fourth day, she ate one-third of what was left. On the fifth day, she ate one-half of what was left. On the sixth day, she ate the one remaining cookie. How many cookies did Bertha's mother bake all together?

A Common Approach

The initial reaction to this problem is to begin to form a series of expressions to represent the number of cookies eaten each day. Let x represent the number of cookies Bertha began with.

Day #	Had	Amount Eaten	Amount Remaining
1	x	$\frac{x}{2}$	$\frac{x}{2}$
2	$\frac{x}{2}$	$\frac{x}{4}$	$\frac{x}{4}$
3	$\frac{x}{4}$	$\frac{x}{16}$	$\frac{3x}{16}$
4	$\frac{3x}{16}$	$\frac{3x}{48}\left(=\frac{x}{16}\right)$	$\frac{x}{8}$
5	$\frac{x}{8}$	$\frac{x}{16}$	$\frac{x}{16}$
6	$\frac{x}{16}$	1	

$$\text{Thus } \frac{x}{16} = 1$$

$$x = 16.$$

She started with 16 cookies.

An Exemplary Solution

A more efficient approach is to use our strategy of working backwards. We begin at the end of the problem, and work backwards to the beginning:

On Day 6, she ate the final 1 cookie; there must have been 1
On Day 5, she ate 1/2, so there must have been 2
On Day 4, she ate 1/3, so there must have been 3
On Day 3, she ate 1/4, so there must have been 4
On Day 2, she ate 1/2, so there must have been 8
On Day 1, she ate 1/2, so there must have been 16.

She started with 16 cookies. Notice that when we work backwards, we must "reverse" the operation we are using. Instead of taking half, we double; instead of adding, we subtract and so on. This appears to be an easier process.

PROBLEM 3.5

A problem that has vexed many mathematics recreationalists is as follows: Maria is 24 years old. Maria is twice as old as Anna was when Maria was as old as Anna is now. How old is Anna?

A Common Approach

The solution to this problem does not lend itself to simply setting up an equation that will lead us to an answer. More is involved. We can begin by setting up Figure 3.4:

	Previously	Currently
Anna	A	$a + x$
Maria	$24 - x$	24

Figure 3.4

We have $24 = 2a$, therefore, $a = 12$. Also $24 - x = a + x = 12 + x$; therefore, $x = 6$. Anna was 12, when Maria was as old (18), as Anna is currently (18).

An Exemplary Solution

Working backwards might prove to be a reasonable way to approach this problem. Therefore, we could also have proceeded as follows:

The situation presented is manifested at two levels:

1. At the current time, when Maria is 24 years old, and
2. At a time n years ago.

We then set up the following relationships:

M = Maria's age (= 24), A = Anna's age, n = the difference between the two time periods.

From the first part: she is twice as old as Anna was:

$$2(A - n) = M. \tag{3.1}$$

From the second part: as Maria as old was as Anna is now follows:

$$M - n = A. \tag{3.2}$$

Equation (3.2) is now substituted into Equation (3.1)

$$2(M - n - n) = M \Rightarrow n = \frac{M}{4} = \frac{24}{4} = 6. \tag{3.3}$$

The value of $n = 6$ inserted into Equation (3.2), yields:

$$M - 6 = A \Rightarrow A = 24 - 6 = 18. \tag{3.4}$$

This tells us that Anna is 18 years old.

PROBLEM 3.6

Which point in a convex quadrilateral is such that the sum of the distances from that point to each of the vertices is a minimum?

A Common Approach

Without much thinking most would attempt a trial and error method of finding the point that would be located so that the sum of the distances to the vertices is the smallest possible. It is possible that one might "stumble" onto the point which is the intersection of the diagonals. This was the correct answer, but this method would not leave the solution without question.

An Exemplary Solution

Our strategy of working backwards would prove to be a rather clever tactic. We begin with quadrilateral *ABCD*, where the diagonals intersect at point *E*, and a point *P*, which we believe might be our desired point of minimum sum distance to each of the vertices. We then draw the (dashed) lines joining point *P* to each of the vertices of the quadrilateral as shown in Figure 3.5.

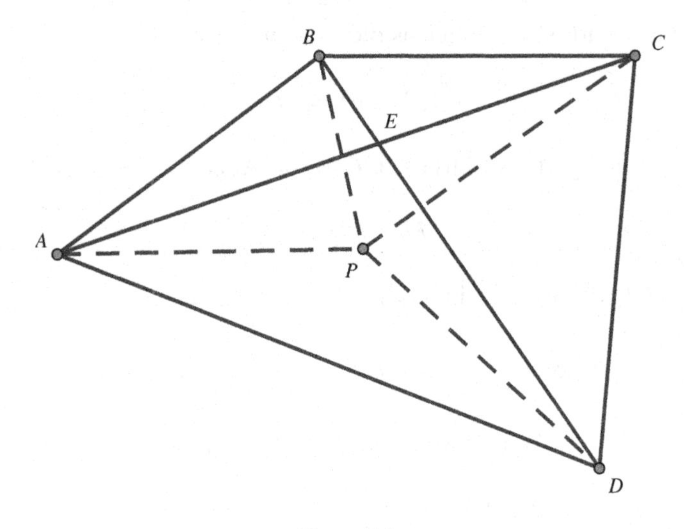

Figure 3.5

As we look at triangle APC we find that $AP + PC > AC$, because the sum of any two sides of a triangle is always greater than the third side. Similarly, $BP + PD > BD$. By addition, we get that $AP + PC + BP + PD > AC + BD$. Therefore, by working backwards we conjectured that P might be our desired point, we find that had we chosen any other point as starting position, the same would have had the same result. The only point, therefore, which satisfies the condition is point E, which is the point of intersection of the diagonals.

PROBLEM 3.7

Suppose the roots of the equation $x^2 + 3x - 3 = 0$ are r and s. What would be the value of $r^2 + s^2$?

A Common Approach

The common approach is to actually solve the equation for the values of r and s. Using the quadratic formula, which is $x = \frac{-b \pm \sqrt{b^2-4ac}}{2a}$, for $ax^2 + bx + c = 0$, we obtain:

$$x = \frac{3 \pm \sqrt{9 - 4 \cdot 1 \cdot (-3)}}{2} = \frac{-3 \pm \sqrt{21}}{2}.$$

We now need to find the squares of these roots and then their sum:

$$r^2 = \frac{15 - \sqrt{21}}{2}$$
$$s^2 = \frac{15 + \sqrt{21}}{2}$$
$$r^2 + s^2 = 15.$$

An Exemplary Solution

To employ a more elegant solution we need to recall a relationship from elementary algebra that states that the sum of the roots of the quadratic equation $ax^2 + bx + c = 0$ is $\frac{-b}{a}$, and the product of the roots is $\frac{c}{a}$. From our given equation we find that the sum of the roots, $r + s = -3$, and the product of the roots, rs is -3. Using our strategy of working backwards, that is, to search for the sum of the squares of the roots rather than to go straight through, as we did earlier, to find the roots we would look to see how we could generate that sum, since $(r + s)^2 = r^2 + s^2 + 2rs$. Rewrite this equation as follows $r^2 + s^2 = (r + s)^2 - 2rs$.

Therefore, the value of $r^2 + s^2 = (-3)^2 - 2(-3) = 9 + 6 = 15$.

PROBLEM 3.8

Max, Sam and Jack are playing an unusual card game. In this game, when a player loses, he gives each of the other players as much money as they have. Max loses the first game and gives Sam and Jack as much money as they each have. Sam loses the second game, and gives Max and Jack as much money as they each have. Jack loses game number three, and gives Max and Sam as much money as they each have. They then decide to stop playing and each has exactly $8.00. How much money did each of them start with?

A Common Approach

The problem suggests we set up a series of equations designed to represent each game. We shall begin by representing the starting money for each as follows:

Max starts with x, Sam starts with y and Jack starts with z,

Game #	Max	Sam	Jack
1	$x - y - z$	$2y$	$2z$
2	$2x - 2y - 2z$	$3y - x - z$	$4z$
3	$4x - 4y - 4z$	$6y - 2x - 2z$	$7z - x - y$

From the last transaction, we find that each of these values is 8. This gives us the following three equations in three variables:

$$4x - 4y - 4z = 8 \quad \text{or} \quad x - y - z = 2$$
$$-2x + 6y - 2z = 8 \quad \text{or} \quad -x + 3y - z = 4$$
$$-x - y + 7z = 8 \quad \text{or} \quad -x - y + 7z = 8.$$

When we solve these three equations simultaneously we obtain:

$$x = 13, \quad y = 7 \quad \text{and} \quad z = 4.$$

An Exemplary Solution

Notice that the problem gave the end situation and asked for the starting situation. This might give us a clue to a problem that will usually benefit from a strategy of working backwards. Notice, too, that the statement of the situation shows that the same amount of money (namely, $24) is always "in play". Working backwards will provide an elegant solution.

	Max	Sam	Jack	
Game # 3	8	8	8	total: 24
Game # 2	4	4	16	total: 24
Game # 1	2	14	8	total: 24
Start:	13	7	4	total: 24

Max started with $13; Sam started with $7; Jack started with $4, the same answers as before, but found in a more elegant fashion.

PROBLEM 3.9

Al and Steve are separating spotted salamanders for the nature museum. Al is putting the ones with 2 spots into one exhibit, while Steve is putting the ones with 7 spots into another. Al has 5 more salamanders in his exhibit than Steve has in his exhibit. Altogether they have salamanders with a total of 100 spots. How many salamanders are in the two exhibits?

A Common Approach

The nature of this problem usually leads the potential problem solver to use algebra. One then begins by letting x represent the number of salamanders that Al has in his exhibit, and y is the number of salamanders that Steve has in his. This gives us the following equations:

$$x - y = 5,$$
$$2x + 7y = 100.$$

The method of solving this pair of simultaneous equations is as follows. Multiplying the first equation by 2 we get:

$$2x - 2y = 10,$$
$$2x + 7y = 100.$$

Subtracting the two equations gives us:

$$9y = 90, \quad \text{or} \quad y = 10.$$

Then by substituting the value of y in the first equation we get $x = 15$. Therefore, Al and Steve together have a total of $15 + 10 = 25$ salamanders. This is a perfectly good solution, but not the most elegant.

An Exemplary Solution

Let us see if we can simplify our work by solving the problem by working backwards. We are not asked how many salamanders each of the two men has. Rather, we are asked for the sum of their salamanders. So, we can start with the same two equations. In other words we seek $x + y$, rather than each of the variables separately. Once again, we will set up the two equations directly from the given information.

$$x - y = 5,$$
$$2x + 7y = 100.$$

However, this time we will look to craft a way to get the sum of the two variables.

To do this, we multiply the first equation by 5 and the second by 2 to get the following.

$$5x - 5y = 25,$$
$$4x + 14y = 200.$$

Then by adding the two equations, we get $9x+9y = 225$ and $x+y = 25$. This method is not typical, but does demonstrate a more sophisticated way of solving problems that ask for something other than what most people expect. Hence, we used a slightly more unusual method of solution.

PROBLEM 3.10

Given the following two equations, we seek to find the sum of $x + y$:

$$6x + 7y = 2007,$$
$$7x + 6y = 7002.$$

A Common Approach

The traditional approach to solving two equations in two variables is to solve them simultaneously.

$$6x + 7y = 2007,$$
$$7x + 6y = 7002.$$

Multiply the first equation by 7 and the second by 6 gives us:

$$42x + 49y = 14049,$$
$$42x + 36y = 42012.$$

Subtracting the two equations, we get:

$$13y = -27963.$$

And then $y = -2151$.

By substituting this value of y into the first equation, we get:

$$6x - 15057 = 2007,$$
$$6x = 17064,$$
$$x = 2844.$$

Therefore, the required sum $x + y = 2844 - 2151 = 693$.

An Exemplary Solution

Let us approach this problem by working backwards. A glance at the two equations that we were given reveals a certain symmetry. We might ask ourselves if that symmetry could lead us to a more elegant solution. Looking at what we were asked to find, we notice that unlike the usual question of this sort we were not asked to find the individual values of x and of y, rather only the value of their sum. So, let us see if the symmetry above could lead us to finding the sum without first seeking the x and y values. If we add the two equations, we get:

$$\begin{array}{r} 6x + 7y = 2007 \\ 7x + 6y = 7002 \\ \hline 13x + 13y = 9009 \end{array}$$

Dividing both sides of the equation by 13, results in $x + y = 693$, and we obtained our desired result by working backwards from what was being asked.

Chapter 4

Adopting a Different Point of View

Of the many strategies that exist and that are available to us to solve mathematical problems, the one that allows us to avoid "running into a brick wall" — namely avoiding frustration — is that of approaching the problem from a different point of view. Perhaps one of the classic examples — because of its simplicity and dramatic difference in solution method — is the following. This is an example where the common method leads to a correct answer, but is cumbersome and can often provide an opportunity for some arithmetic mistakes. As an example, consider the following problem:

At a school with 25 classes, each of these sets up a basketball team to compete in a school-wide tournament. In this tournament a team that loses one game is immediately eliminated. The school only has one gymnasium, and the principal of the school would like to know how many games will be played in this gymnasium in order to get a winner.

The typical solution to this problem could be to simulate the actual tournament by beginning with 12 randomly-selected teams playing a second group of 12 teams — with one team drawing a bye — that is, passing up a game. This would then continue with the winning teams playing each other as shown here.

Any **12 teams** versus any other **12 teams,** which leaves **12 winning teams** in the tournament.

6 winners versus **6 other winners**, which leaves **6 winning teams** in tournament.

3 winners versus **3 other winners,** which leaves **3 winning teams** in tournament.

3 winners + **1 team** (which drew a bye) = **4 teams**.

2 remaining **teams** versus **2** remaining **teams**, which leaves **2 winning teams** in the tournament.

1 team versus **1 team** to get a **champion**!

Now counting up the number of games that have been played we get:

Teams playing	**Games played**	Winners
24	**12**	12
12	**6**	6
6	**3**	3
3+ 1 bye = 4	**2**	2
2	**1**	1

The total number of games played is:

$$12 + 6 + 3 + 2 + 1 = 24.$$

This seems like a perfectly reasonable method of solution, and certainly a correct one.

Approaching this problem from a different point of view would be vastly easier by considering the losers, rather than winners, which is what we did in the previous solution. In this case, we ask ourselves, how many losers must there have been in this competition in order to get one champion? Clearly, there had to be 24 losers. To get 24 losers, there needed to be 24 games played. And with that the problem is solved. Looking at the problem from an alternative point of view is a curious approach that can be useful in a variety of contexts.

Another alternative point of view would be to consider these 25 teams with one of them — only for our purposes — considered to be a professional basketball team that would be guaranteed to win the tournament. Each of the remaining 24 teams would be playing the professional team only to lose. Once again, we see that 24 games are required to get a champion. This should demonstrate for you the power of this problem-solving technique. We now consider a wide variety of problems that can be most efficiently solved by adopting a different point of view.

PROBLEM 4.1

A point P is selected on the circumference of a circle, O (Figure 4.1). PA and PB are drawn perpendicular to two mutually perpendicular diameters. If $AB = 12$, what is the area of the circle in terms of π?

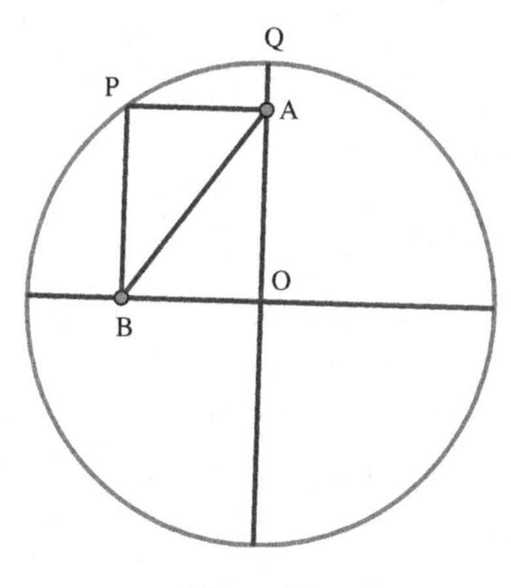

Figure 4.1

A Common Approach

Most people approach the problem by trying to make use of the Pythagorean Theorem, since triangles PAB and OAB are right triangles. This approach leads to a dead end, since there is not enough information given to apply the Pythagorean Theorem properly.

An Exemplary Solution

This problem can be done in a number of ways. One way is to consider extremes. Suppose point P is selected on the circle at point Q. In that case AB would overlap QO, which is the radius of the circle. Therefore, the area of the circle would be 144π.

We can also look at this problem from another point of view. A quadrilateral with three right angles must be a rectangle. The line segment AB is a diagonal of this rectangle. Similarly, PO is also a diagonal of the rectangle. Since the diagonals of rectangle are equal, the radius of the circle, $PO = 12$, and the area is therefore again 144π.

PROBLEM 4.2

A standard deck of 52 playing cards is randomly split into two piles with 26 cards in each. How does the number of red cards in one pile compare to the number of black cards in the other pile?

A Common Approach

The typical way that one approaches this problem is to represent the number of black cards and the number of red cards in each pile in a symbolic fashion: We can represent the situation symbolically as follows:

$$\begin{aligned} \mathrm{B1} &= \text{the number of black cards in pile \#1}, \\ \mathrm{B2} &= \text{the number of black cards in pile \#2}, \\ \mathrm{R1} &= \text{the number of red cards in pile \#1}, \\ \mathrm{R2} &= \text{the number of red cards in pile \#2}. \end{aligned}$$

Then, since the total number of black cards equals 26, we can write this as $\mathrm{B1} + \mathrm{B2} = 26$, and since the total number of cards in pile #2 equals 26, we have $\mathrm{R2} + \mathrm{B2} = 26$.

By subtracting these two equations: $\mathrm{B1}+\mathrm{B2} = 26$, and $\mathrm{R2}+\mathrm{B2} = 26$, we get: $\mathrm{B1} - \mathrm{R2} = 0$. Therefore, we have $\mathrm{B1} = \mathrm{R2}$, which tells us that the number of red cards in one pile equals the number of black cards in the other pile. Although this solves the problem, there is nothing elegant about the solution. Our theme in this chapter is to provide clever solutions to show the beauty and power in mathematics.

An Exemplary Solution

Perhaps as a more clever alternative approach, we shall take all the red cards in pile #1 and switch them with the black cards in pile #2. Now, all the black cards will be in one pile, and the red cards in the other pile. Therefore, the number of red cards in one pile and the number of black cards in the other pile had to be equal to begin with. Simple logic solves the problem — just looking at the problem from another point of view.

PROBLEM 4.3

Lohengrin was given four pieces of chain (Figure 4.2), each consisting of three links (shown below). Show how these four pieces of chain could be made into a circular chain by opening and closing, *at most*, three links.

Figure 4.2

A Common Approach

Typically, a first attempt at a solution involves opening the end link of one chain, joining it to the second chain to form a 6-link chain; then opening and closing a link in the third chain and joining it to the 6-link chain to form a nine-link chain. By opening and closing a link in the fourth chain and joining it to the 9-link chain, a 12-link chain, *which is not a circle* is obtained. Thus, this traditional attempt usually ends unsuccessfully. Some folks typically try other combinations of open/closing one link of each of various chains pieces to try to join them together to get the desired result, but this approach will not work.

An Exemplary Solution

This problem lends itself quite nicely to the strategy of adopting a different point of view. Actually one might say that it proves to be invaluable. Instead of continually trying to open and close ***one*** *link of each chain piece*, a different point of view would involve opening ***all*** *the links in one chain* and using those links to connect the remaining three chain pieces together, and into the required circle chain. This quickly gives the successful conclusion.

PROBLEM 4.4

Which natural numbers less than 100 leave a remainder of 3 when divided by 7 and a remainder of 4 when divided by 5?

A Common Approach

Let us consider the set of natural numbers less than 100 leaving the remainder of 3 on division by 7, which are as follows: $\{3, 10, 17, 24, 31, 38, 45, 52, 59, 66, 73, 80, 87, 94\}$. Now we will consider the set of natural numbers less than 100 which leave a remainder of 4 when divided by 5: $\{4, 9, 14, 19, 24, 29, 34, 39, 44, 49, 54, 59, 64, 69, 74, 79, 84, 89, 94, 99\}$.

When we inspect these 2 sets we noticed there were 3 numbers common to both sets. They are 24, 59, and 94.

An Exemplary Solution

We will consider this problem from another point of view. Each of the numbers that we seek must be of the form $7n + 3$, and also of the form $5k+4$, where n and k are integers. We can combine these properties so that we are seeking numbers of the form $35r + p$, where r and p are integers. The first set of numbers which are the form $7n + 3$ can also be written as $35r + 3$, $35r + 10$, $35r + 17$, $35r + 24$, and $35r + 31$. Only one of these is also the form $5k + 4$, namely, $35r + 24$. Considering the numbers less than 100 satisfying this relationship, we set $r = 0$, 1, and 2 to get 3 required numbers: 24, 59, and 94.

PROBLEM 4.5

Which of the following two expressions is greater? $\sqrt{5} + \sqrt{8}$ or $\sqrt{4} + \sqrt{10}$

A Common Approach

With the ubiquity of calculators it is no wonder that one would take the square root of each of the numbers and then calculate their sums to get the required answer. Although this could be relatively efficient, it is not what we would consider an elegant solution.

An Exemplary Solution

We will approach this problem from another point of view, namely, to take the square of each of these sums and see if the comparison will be then forthcoming.

$$\left(\sqrt{5}+\sqrt{8}\right)^2 = 5+2\sqrt{40}+8 = 13+2\sqrt{40}$$
$$\left(\sqrt{4}+\sqrt{10}\right)^2 = 4+2\sqrt{40}+10 = 14+2\sqrt{40}.$$

Having simplified it this way, the answer is now obvious that $\sqrt{4}+\sqrt{10}$ is the greater of the two expressions.

PROBLEM 4.6

What are all the positive integral values of n for which the fraction $\frac{7n+15}{n-3}$ is also an integer?

A Common Approach

An immediate response is to try various values of n and see which result in integers. For instance, if we let $n = 4$, we obtain $\frac{43}{1}$ which is integral. While this approach might yield some of the values of n, how do we know whether we have found *all* of them? This approach will usually not give all the possibilities.

An Exemplary Solution

Let us use our strategy of adopting a different point of view. We begin by actually performing the division:

$$\begin{aligned}\frac{7n+15}{n-3} &= \frac{7n-21+36}{n-3} \\ &= \frac{7n-21}{n-3}+\frac{36}{n-3} \\ &= \frac{7(n-3)}{n-3}+\frac{36}{n-3} = 7+\frac{36}{n-3}.\end{aligned}$$

But for this value to be an integer $n - 3$ must be a factor of 36. The factors of 36 are 1, 2, 3, 4, 6, 9 12, 18 and 36. Therefore,

If $n - 3 =$	then $n =$
1	4
2	5
3	6
4	7
6	9
9	12
12	15
18	21
36	39

The values of n for which the fraction $\frac{7n+15}{n-3}$ is integral are 4, 5, 6, 7, 9, 12, 15, 21, and 39.

PROBLEM 4.7

Each of the 10 court jewelers gave the king's advisor, Mr. Sachs, a stack of gold coins. Each stack contained 10 coins. The real coins weighed exactly 1 ounce each. However, one and only one stack contained "light" coins, each having had exactly 0.1 ounce of gold shaved off the edge. Mr. Sachs wishes to identify the crooked jeweler and the stack of light coins with just one single weighing on a scale. How can he do this?

A Common Approach

The traditional procedure is to begin by selecting one of the stacks at random and weighing it. This trial and error technique offers only a 1 chance in 10 of being correct. Once this is recognized, one may revert to attempt to solve the problem by reasoning. First of all, if all the coins were true, their total weight would be 10×10, or 100 ounces. Each of the 10 counterfeit coins is lighter, so there will be a deficiency of 10×0.1, or 1 ounce. But thinking in terms of the overall deficiency does not lead anywhere, since the 1-ounce shortage will occur whether the counterfeit coins are in the first stack, the second stack, the third stack, etc.

An Exemplary Solution

Let us try to solve the problem by organizing the data in a different fashion. We must find a method for varying the deficiency in a way that permits us to identify the stack from which the counterfeit coins are taken. Label the stacks #1, #2, #3, #4, . . . , #9, #10. Then we take one coin from stack #1, two coins from stack #2, three coins from stack #3, four coins from stack #4, etc. We now have a total of $1 + 2 + 3 + 4 + \cdots + 8 + 9 + 10 = 55$ coins. If they were all true, the total weight would be 55 ounces. If the deficiency were 0.5 ounces, then there were 5 light coins, taken from stack #5. If the deficiency were 0.7 ounces, then there were 7 light coins, taken from stack #7 and so on. Thus, Mr. Sachs could readily identify the stack of light coins, and consequently the jeweler who had shaved each coin.

PROBLEM 4.8

A fast food restaurant sells chicken nuggets in boxes of 7 and 3. What is the greatest number of nuggets a person *cannot* buy?

A Common Approach

We simply try to find the answer by actually making combinations of 7 and 3 until we reach a point where we can make all numbers of nuggets.

Number of Nuggets		
1	no	
2	no	
3	yes	1×3
4	no	
5	no	

(Continued)

(*Continued*)

Number of Nuggets		
6	yes	3×3
7	yes	1×7
8	no	
9	yes	3×3
10	yes	1×7 and 1×3
11	no	
12	yes	4×3
13	yes	2×3 and 1×7
14	yes	2×7
15	yes	5×3
16	yes	3×3 and 1×7
17	yes	1×3 and 2×7
18	yes	6×3
19	yes	4×3 and 1×7
20	yes	2×3 and 2×7
21	yes	7×3 or 3×7
22	yes	5×3 and 1×7

It appears that the largest number of nuggets we cannot buy is 11. From now on, all we need do is add another 3 or another 7.

An Exemplary Solution

Here, we will invoke a concept in mathematics that will show some elegance and will leave the reader to wonder why it is true — a motivation for some further investigation. There is a famous theorem known as the "Chicken McNuggets Theorem". The theorem states that if McDonald's sells Chicken McNuggets in boxes of a or b McNuggets, where a and b have no common factor, then the largest number of McNuggets one *cannot* buy is $ab - (a+b)$. For example, if they are sold in boxes of 8 and 5, then the greatest number that *cannot* be sold is $8 \cdot 5 - (8 + 5) = 40 - 13 = 27$.

In our problem (above), the greatest number that cannot be sold is $3 \cdot 7 - (3 + 7)$ or $21 - 10 = 11$.

PROBLEM 4.9

Simplify each of the following:

(a)
$$\frac{729^{35} - 81^{52}}{27^{69}},$$

(b)
$$\frac{6 \cdot 27^{12} + 2 \cdot 81^{9}}{8000000^{2}} \cdot \frac{80 \cdot 32^{3} \cdot 125^{4}}{9^{19} - 729^{6}}.$$

A Common Approach

Although one may be tempted to use a calculator to evaluate his expression, all too often our expectations for the calculator are overestimated and the result comes back with an "error" message.

An Exemplary Solution

We will approach this problem from another point of view. With the knowledge of powers of 3, the problem can be rather cleverly evaluated as follows:

(a)
$$\frac{729^{35} - 81^{52}}{27^{69}} = \frac{(3^6)^{35} - (3^4)^{52}}{(3^3)^{69}}$$
$$= \frac{3^{210} - 3^{208}}{3^{207}} = \frac{3^{208} \cdot (3^2 - 1)}{3^{207}} = 3 \cdot 8 = 24.$$

This expression can be simplified by breaking the numbers up into prime factors as follows:

(b)
$$\frac{6 \cdot 27^{12} + 2 \cdot 81^{9}}{8000000^{2}} \cdot \frac{80 \cdot 32^{3} \cdot 125^{4}}{9^{19} - 729^{6}}$$
$$= \frac{2 \cdot 3 \cdot (3^3)^{12} + 2 \cdot (3^4)^{9}}{(3^2)^{19} - (3^6)^{6}} \cdot \frac{2^4 \cdot 5 \cdot (2^5)^3 \cdot (5^3)^4}{(2^3 \cdot 2^6 \cdot 5^6)^2}$$
$$= \frac{2 \cdot 3^{37} + 2 \cdot 3^{36}}{3^{38} - 3^{36}} \cdot \frac{2^{19} \cdot 5^{13}}{2^{18} \cdot 5^{12}} = \frac{2 \cdot 3^{36}(3+1)}{3^{36}(3^2 - 1)} \cdot 2 \cdot 5$$
$$= \frac{2(3+1)}{3^2 - 1} \cdot 2 \cdot 5 = 10.$$

PROBLEM 4.10

Wolfgang and Ludwig each has a whole number of Euros, each has fewer than 100 Euros. When they count their money, it turns out that three-fourths of Wolfgang's money equals two-thirds of Ludwig's money. What is the maximum number of Euros each of them can have?

A Common Approach

The first reaction is to use algebra. We can set up one equation in two variables. Let W represent the number of Euros Wolfgang has, and let L represent the number of Euros Ludwig has. We can then set up the following equation:

$$\frac{3W}{4} = \frac{2L}{3}.$$

Multiplying by 12, we get: $9W = 8L$. When we solve for W, we have:

$$W = \frac{8L}{9}.$$

Since they each have a whole number of Euros, Ludwig must have a number of Euros that is a multiple of 9, such as $9, 18, 27, 36, \ldots, 99$ Euros. We can now try each of these in turn to find the number of Euros Ludwig has. The largest number of Euros Ludwig can have is 11×9 or 99 Euros (less than 100). Since $\frac{2}{3}$ of Ludwig's money (66 Euros) is $\frac{3}{4}$ of Wolfgang's money. Wolfgang has $\frac{4}{3} \times 66$ or 88 Euros, while Ludwig has 99 Euros.

An Exemplary Solution

Let us use arithmetic and adopt a different point of view. Since $\frac{3}{4}$ of Wolfgang's money equals $\frac{2}{3}$ of Ludwig's money, we shall look for equivalent fractions having the same numerator:

$$\text{Wolfgang: } \frac{3}{4} = \left[\frac{6}{8}\right] = \frac{9}{12},$$

$$\text{Ludwig: } \frac{2}{3} = \frac{4}{6} = \left[\frac{6}{9}\right].$$

If Wolfgang had 8 Euros and Ludwig had 9 Euros, then the fractional parts would be equal, namely 6 Euros each. So our answers must be the same multiplier of both 8 and 9. Thus, the greatest amount of Euros Ludwig can have is 11×9 or 99 Euros, while this gives Wolfgang 11×8 or 88 Euros.

We can check these answers by taking $\frac{3}{4}$ of 88 Euros, which is 66, and taking $\frac{2}{3}$ of 99 Euros also equals 66.

PROBLEM 4.11

In the Figure 4.3, the dimensions of rectangle $AEFK$ are given as width $AK = 8$, while the length AE is divided into four segments such that $AB = 1, BC = 6, CD = 4$ and $DE = 2$. What is the area of the four shaded triangles?

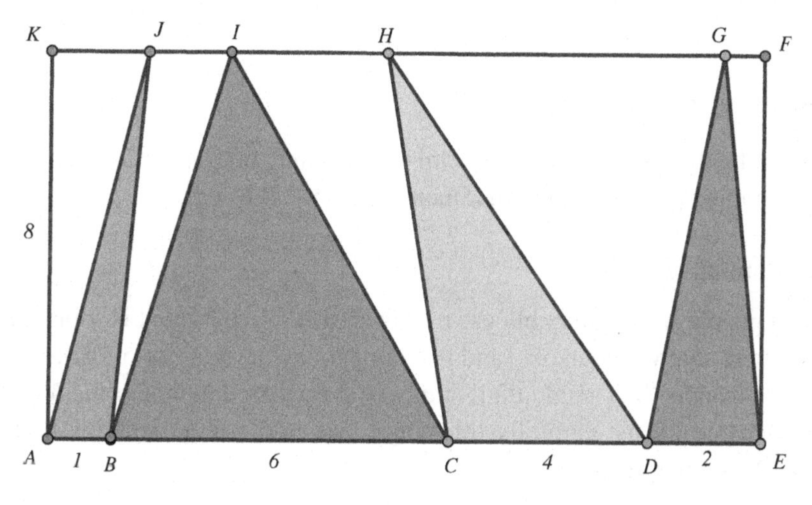

Figure 4.3

A Common Approach

The obvious approach is to find the area of each of the four triangles and get their sum. In all four cases, the altitude of the triangle equals the length of $AK = 8$. Therefore, the areas of the four triangles are:

$$\Delta ABJ = \frac{1}{2} \cdot 1 \cdot 8 = 4$$

$$\Delta BCI = \frac{1}{2} \cdot 6 \cdot 8 = 24$$

$$\Delta CDH = \frac{1}{2} \cdot 4 \cdot 8 = 16$$

$$\Delta DEG = \frac{1}{2} \cdot 2 \cdot 8 = 8$$

The sum of these areas is $4 + 24 + 16 + 8 = 52$ square units.

An Exemplary Solution

We can make use of our strategy of adopting a different point of view to solve this problem. The triangles each have the same altitude, namely, 8. The sum of the bases of the four triangles equals the length of the longer side of the rectangle, which is 13. Thus, the area of the four shaded triangles is half the area of the rectangle or $\frac{1}{2} \cdot 13 \cdot 8 = 52$.

PROBLEM 4.12

Using the digits from 1 to 9 determine how many numbers can be formed so that the digits of that number increase from left to right?

A Common Approach

Most people would probably use trial and error to see if there is some kind of pattern that would evolve and list numbers by the number of digits they have — that is, first single-digit numbers, then 2-digit numbers and 3-digit numbers etc. Done carefully this could lead to a correct solution, but it would be rather tedious.

An Exemplary Solution

Let us first consider the set of integers that we have at our disposal $\{1, 2, 3, 4, 5, 6, 7, 8, 9\}$. Every subset of these numbers except the empty set would produce one of our desired numbers. For example, the set $\{3, 5, 7, 9\}$ would give us the required number 3579. The question, then, is, how many subsets are there of this set of nine digits? There are $2^9 = 512$ such subsets. However, this number includes the empty set, which we must deduct. Therefore there are $2^9 - 1 = 511$ subsets of our 9 digits each of which gives us a number, which can satisfy our requirement of having its digits in ascending order from left to right.

PROBLEM 4.13

In Figure 4.4, we have an isosceles triangle with an infinite series of circles, each of which is tangent to the 2 equal sides of the isosceles triangle and to the adjacent circles, with the bottom circle tangent to the base of the triangle. The sides of the isosceles triangle are 13, 13 and 10. What is the sum of the circumferences of these circles?

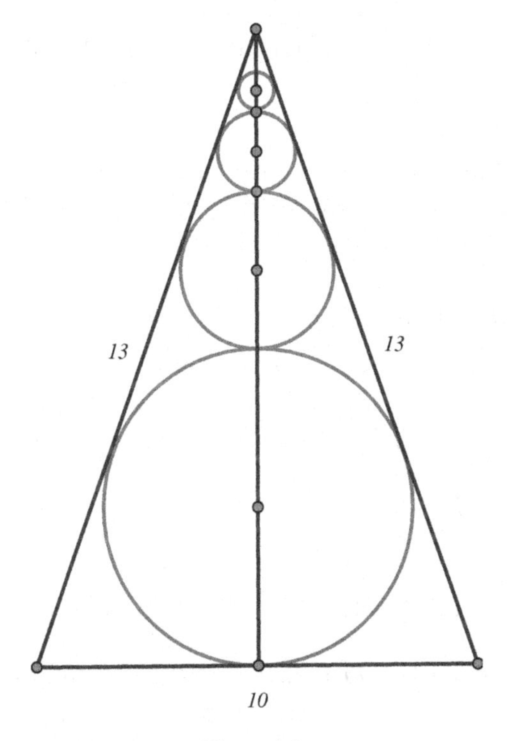

Figure 4.4

A Common Approach

As tedious as it sounds a common approach here would be to find the circumference of each of the circles and then take their sum. This would be a very complicated computation, but done carefully could lead to a correct solution.

An Exemplary Solution

We will use the strategy where we will consider this problem from another point of view. Using the Pythagorean Theorem we find that the altitude of this isosceles triangle has length 12. We will notice that the sum of the diameters is equal to the altitude of the isosceles triangle, since there is an *infinite* number of circles. Therefore, the sum of the circumferences of the circles is the sum of the diameters times π, which is 12π.

PROBLEM 4.14

What is the smallest nonnegative remainder when 22^7 is divided by 123?

A Common Approach

Unfortunately, the common way of approaching this problem would be to actually spend a lot of time evaluating the large number 22^7 and dividing it by 123 to see what the remainder would be.

An Exemplary Solution

We will consider this problem from another of the point of view. Rather than to expand 22^7 as a number without an exponent, we will expand it in terms of exponential powers. That is, we can write 22^7 as follows:

$$\begin{aligned} 22^7 &= (2^7)(11^7) \\ &= (2^7)(11^2)(11^2)(11^2)(11) \\ &= (123+5)(123-2)(123-2)(123-2)(11). \end{aligned}$$

Now we need to recall that when we have two binomials such as, for example, $123 + s$ and $123 + t$ their product can be shown to be equal to $123k + st$ in the following way:

$$\begin{aligned} (123+s)(123+t) &= 123^2 + 123s + 123t + st \\ &= 123(123+s+t) + st = 123k + st. \end{aligned}$$

Therefore, from the above, we have

$$123n - 440 = 123n - 492 + 52 = 123(n-4) + 52.$$

Thus we have a remainder of 52 when 22^7 is divided by 123.

PROBLEM 4.15

In a football game, teams receive 2 points for a safety, 3 points for a field goal, and 7 points for a touchdown. If we eliminate the 2 points for a safety, only the 3-point score and the 7-point score are left. What is the highest score that cannot be made in this game?

A Common Approach

The obvious approach is to write out all the possible scores until we are certain that there are no higher scores that can be made. Yet, how can we be certain no higher score exists?

An Exemplary Solution

We can use our strategy of adopting a different point of view to solve the problem. Instead of looking at which scores *cannot* be achieved, let's look at those that *can* be made. The scores that can be made with a field goal are 3, 6, 9, 12, 15, ... The scores that can be made with a touchdown are 7, 14, 21, 28, ... Other scores can be made by adding either a field goal or a touchdown to a previous score. Thus we cannot score any of these 2, 4, 5, 8, 11. Any score beginning with 12 can be made, as we can see from the following:

$\mathbf{12} = 4{\cdot}3$	$\mathbf{15} = 5{\cdot}3$	$\mathbf{18} = 6{\cdot}3$
$\mathbf{13} = (2{\cdot}3) + (1{\cdot}7)$	$\mathbf{16} = (3{\cdot}3) + (1{\cdot}7)$	$\mathbf{19} = (4{\cdot}3) + (1{\cdot}7)$
$\mathbf{14} = 2{\cdot}7$	$\mathbf{17} = (1{\cdot}3) + (2{\cdot}3)$	$\mathbf{20} = (2{\cdot}3) + (2{\cdot}7)$

Thus, the highest score that cannot be made is 11.

It is interesting to note that there is a purely mathematical theorem that covers this situation.

Given two relatively prime numbers (a and b), the highest score that cannot be made is their product minus their sum or $(a \cdot b) - (a + b)$. In this case, that would be $(7 \cdot 3) - (7 + 3)$ or $21 - 10 = 11$.

PROBLEM 4.16

The number 6! (reads as "six factorial") equals the product of $6 \cdot 5 \cdot 4 \cdot 3 \cdot 2 \cdot 1 = 720$. What is the value of $\frac{100!-99!-98!}{100!+99!+98!}$?

A Common Approach

The most obvious approach is to write out all of the factorial expressions, use a calculator or computer, and actually compute the results. This would give the answer, but it requires a lot of cumbersome arithmetic.

An Exemplary Solution

Let us use our strategy of adopting a different point of view. Each of the factorials can be written with a common factor of 98! Thus, we can write 100! as $100 \cdot 99 \cdot 98!$ and 99! as $99 \cdot 98!$, Following this approach, we get:

$$\frac{100! - 99! - 98!}{100! + 99! + 98!} = \frac{98!(100 \cdot 99 - 99 - 1)}{98!(100 \cdot 99 + 99 + 1)}$$

$$= \frac{(100 \cdot 99 - 99 - 1)}{(100 \cdot 99 + 99 + 1)} = \frac{9800}{10000} = \frac{49}{50},$$

which is the answer to the original complicated-looking problem.

PROBLEM 4.17

If we divide 450 by an odd number, the resulting quotient is a prime number and there is no remainder. What is the odd number?

A Common Approach

The common approach is to divide 450 by successive odd numbers $(1, 3, 5, \ldots)$ until the resulting quotient is a prime number. This will eventually give the result but may take a long time.

An Exemplary Solution

We can make use of our strategy of adopting a different point of view. The number 450 can be written in simplest form as $2 \cdot 3^2 \cdot 5^2$. Since 3^2 and 5^2 are both odd numbers, and 450 is obviously even, the only possible even prime factor of 450 is 2. Thus, the odd number is $3^2 \cdot 5^2 = 225$.

PROBLEM 4.18

The number 1,000,000 has many pairs of integer factors — that is two numbers whose product is 1,000,000. Yet, there is only one such pair of factors that does not contain any zeros. What are these factors of 1,000,000?

A Common Approach

The traditional approach is to try various pairs of numbers whose product is 1,000,000, to search for the pair that contains no digit 0. We can begin with $1 \times 1{,}000{,}000$, $2 \times 500{,}000$ and so on. This can surely take a great deal of time. After all, there are many, many pairs of factors of 1,000,000 to try.

An Exemplary Solution

Let us examine 1,000,000 by adopting a different point of view. The number 1,000,000 can be expressed as 10^6. In turn, this can be expressed as $(2 \times 5)^6 = 2^6 \times 5^6$. This gives us two factors of 1,000,000 that do not contain any zeros are $2^6 = 64$ and $5^6 = 15{,}625$. You will notice that any other pair of factors of 1,000,000 must contain at least one zero, since when the factors of 2 and 5 are combined a multiple of 10 is created and that will produce a zero-ending number.

Chapter 5

Considering Extreme Cases

In order to obtain some help with solving a particular problem, we may consider some variables at their extremes, while holding others constant. If no specificity for the variable has been prescribed, the extreme case might just provide us with some helpful insight. Most of us have used this strategy in real life, without consciously realizing that we were even doing so. We say to ourselves, "What is the worst that can happen?" — for a particular, situation up for a decision. This "worst-case scenario" is an example of using an extreme case that can sometimes help resolve a problem situation in a very clever way. Similarly, suppose you were asked to test a new product, say, laundry soap. You would have to test it in very cold water as well as very hot water — clearly a consideration of two extreme cases that makes the testing worthwhile. If it does well in these two extreme temperatures, it should do well in the temperatures in between.

Sometimes, using extremes to solve a problem could be counterintuitive. For example, when the question arises as to whether it is best to run through a rainstorm to get from point A to point B, or to walk more slowly, one tends to recall that when a car drives fast through a rainstorm the front windshield is overwhelmed with water, while traveling slowly, less water accumulates on the windshield. Does this mean that one should run through the rainstorm, or not? When considering the extreme situations: walking very slowly increases the time in the rainstorm — or to consider the extreme walking at the slowest speed, say, zero — we would get drenched. Therefore, the faster we go, the less wet we would get. Here, using extremes helped solve the problem.

Let us look at a problem where this strategy of considering extremes helps to solve it.

> A series of 40 mailboxes in a local post office receives mail every morning. One day the postmaster distributed 121 pieces of mail to these mailboxes. When he finished, he was surprised to find that one mailbox had more letters than any other. What is the fewest number of letters this mailbox might have had?

Since the problem asked for the *fewest* number of pieces of mail the box could have we can consider the following extreme case. We will spread the mail out as evenly as we can. Suppose every mailbox contained the same number of letters, which is an extreme case, where the opposite extreme would be that one letter box received all the letters. Evenly distributing the letters would be have each box containing 3 letters, since $120 \div 40 = 3$. Then the one remaining letter would give one mailbox 4 letters, which would be the largest number. The smallest number of letters a mailbox could contain and still be more than any other would be 4.

To gain some more practice with this problem-solving technique we will consider another problem — this one with a statistical bent:

> Clarissa wrote down a set of 5 whole numbers. She found out that their mode was 12 and the median number was 14. Their arithmetic mean (or average) was 16. One of the numbers is exactly 5 more than the median. What are the five numbers Clarissa wrote?

Let us use the strategy of considering an extreme case. Since the mode was 12, the worst-case scenario (smallest number of scores) is that there were exactly two 12s. We know that the median or middle score was 14. Since one number was 5 more than the median, one number must have been $14 + 5$ or 19. So far we know the numbers are:

$$12, 12, 14, 19.$$

The mean is found by adding all five numbers and dividing the total by 5. Since the mean score is 16, the total for all 5 numbers must be $16 \times 5 = 80$. So far, we have $12 + 12 + 14 + 19 = 57$. The missing number must be $80 - 57 = 23$. The five numbers Clarissa wrote were 12, 12, 14, 19, 23. Notice how crucial it was to begin the solution with the extreme situation determining that there had to be two 12s.

A word of caution about using this strategy. When we consider an extreme case, we must be careful not to change a variable that affects other variables as well, nor affects a change in the very nature of the problem. The problems presented in this chapter should help identify the situations where this strategy can be employed.

PROBLEM 5.1

A car is driving along a highway at a constant speed of 55 miles per hour. The driver notices a second car, exactly $\frac{1}{2}$ mile behind him. The second car passes the first, exactly 1 minute later. How fast was the second car traveling, assuming its speed is constant?

A Common Approach

The traditional solution is to set up a series of "Rate × Time = Distance" boxes, which many text books guide students to use for this sort of problem. This would be done as follows:

Rate	**× Time =**	**Distance**
55	$\frac{1}{60}$	$\frac{55}{60}$
x	$\frac{1}{60}$	$\frac{x}{60}$

$$\frac{55}{60} + \frac{1}{2} = \frac{x}{60}$$

$$55 + 30 = x$$

$$x = 85.$$

The second car was traveling at a rate of 85 miles per hour.

An Exemplary Solution

An alternate approach would be using the strategy of considering extremes. We assume that the first car is going *extremely* slowly, that is, at 0 miles per hour. Under these conditions, the second car travels $\frac{1}{2}$ mile in one minute to catch the first car. Thus, the second car would have to travel 30 miles per hour. When the first car is moving at 0 miles per hour, then the second car is traveling 30 m.p.h. faster than the first car. If, on the other hand the first car is traveling at 55 m.p.h, then the second car must be traveling at 85 miles per hour (within the legal limit, of course!).

PROBLEM 5.2

We are given parallelograms *ABCD* and *APQR*, with point *P* on side *BC* and point *D* on side *RQ*, as shown in Figure 5.1. If the area of parallelogram *ABCD* is 18, what is the area of parallelogram *APQR*?

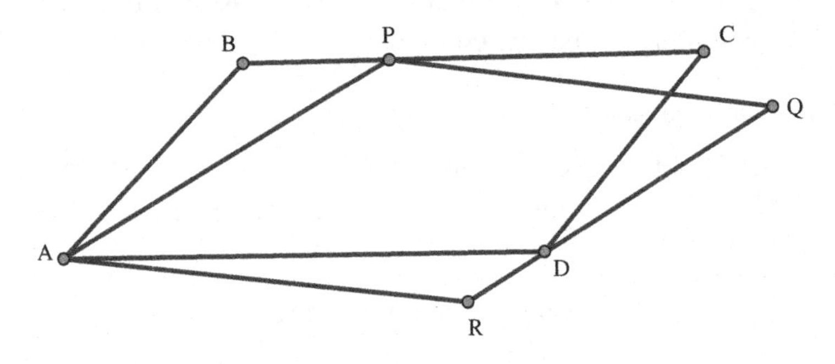

Figure 5.1

A Common Approach

This is by no means an easy problem to solve. First attempts to solve the problem would be to look for congruent relationships that would lead to equal areas. This method will lead nowhere. A clever method, although rather than "off the beaten path" is to draw the line segment *PD*, as shown in Figure 5.2.

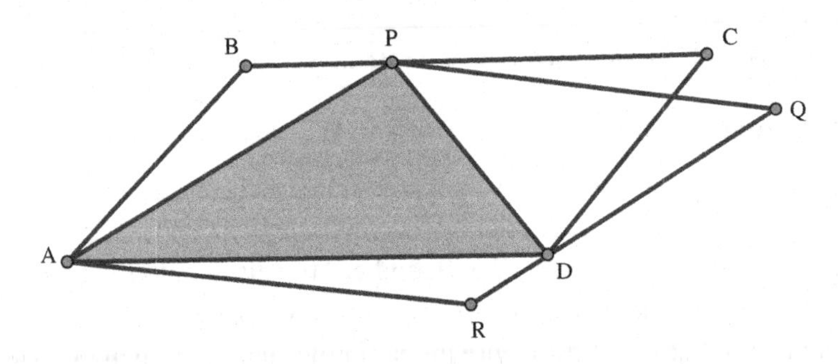

Figure 5.2

Then notice that triangle APD can be shown to be one-half of the area of each of the two parallelograms, since in each case it shares a base with each of the parallelograms as well as the same height. Although this is a rather clever approach to a rather challenging problem, there is yet an even more elegant way to approach this problem.

An Exemplary Solution

When the problem was posed we were merely told that point P was on side BC, but not where along the side it was to be placed. We can consider an extreme case. Therefore, we could have placed P to overlap point B. Similarly, point D, which was to be placed on side RQ, could just as easily have been placed to overlap point R. Under these circumstances, this would certainly fit the original problem's statement, the two parallelograms would overlap, and consequently would have the same area. Therefore, the area of parallelogram $APQR$ is 18.

PROBLEM 5.3

The total distance between Exits 1 and 20 on the new highway is 140 miles. Any two exits must be at least 7 miles apart. What is the maximum distance between any two consecutive exits?

A Common Approach

The usual approach is to try various combinations of numbers hoping to find a maximum. There must be a better way.

An Exemplary Solution

Let us use the strategy of considering extreme cases. First of all, between Exits 1 and 20 there are 19 "distances". Since the minimum distance between any two exits must be 7 miles, suppose we consider the extreme case; namely all the exit distances except for one of them to be 7 miles. Then there will be $18 \times 7 = 126$ miles as the minimum total distances for these 18 "spaces". This leaves a maximum of $140 - 126 = 14$ miles as the maximum distance between any two exits, because if there were 14 miles between any two exits, then there would not be enough miles to allow there to be a 7 mile space between all the other exits.

PROBLEM 5.4

We have two 1-liter bottles. One contains a half-liter of red wine and the other contains a half-liter of white wine. We take a tablespoonful of the red wine and pour it into the white wine bottle, and thoroughly mix the two colored wines. Then we take a tablespoon of this new mixture (red wine and white wine) and pour it into the red wine bottle.

Is there more red wine in the white wine bottle, or more white wine in the red wine bottle?

A Common Approach

There are several common approaches, where the problem solver attempts to solve the problem using the given information such as the tablespoon, which may be extraneous. With some luck and cleverness a correct solution may evolve, but it will not be easy and often not convincing.

An Exemplary Solution

We can see that the size of the spoon does not really matter, since there are large and small tablespoons. Suppose we use a very large tablespoon, one that is enormously large and actually can hold a half liter of liquid — this would be an extreme consideration. When we pour the half liter of the red wine into the white wine bottle, the mixture is then 50% red wine and 50% white wine. After mixing these two together, we take our half-liter spoon and take half of the quantity of this red wine — white wine mixture and pour it back into the red wine bottle. The mixture is now the same in both bottles; so to answer our question, we can conclude that there is as much red wine in the white wine bottle as there is white wine in the red wine bottle.

PROBLEM 5.5

Find the missing digits in the following 7-digit number, so that the number itself is equal to the product of three consecutive numbers. What are these three numbers? 1, 2 _ _, _ _ 6

A Common Approach

One may begin by simply guessing and testing various numbers in the hope that they might get lucky and guess the digits. This is highly unlikely, although possible.

An Exemplary Solution

Instead, let us utilize the strategy of considering extreme cases. The smallest possible number would be 1,200,006 while the largest possible number is 1,299,996. Since we are looking for our answer to be the product of three consecutive numbers, let us examine the cube root of each of these extremes to determine the approximate magnitude of the three numbers.

The cube root of 1,200,006 is approximately 106, while the cube root of 1,299,996 is approximately 109. This limits our choices a great deal. Furthermore, the given number has a units digit of 6. Thus, our three consecutive numbers must end with either 1, 2, and 3, or with 6, 7, and 8, since these are the products whose units digit is 6. With these two clues, our numbers are easily found as 106, 107, and 108. Their product is 1,224,936 and the problem is solved.

PROBLEM 5.6

In Figure 5.3, *ABCD* is a rectangle with sides 8" and 12" as shown. Find the area of the shaded portion of the rectangle.

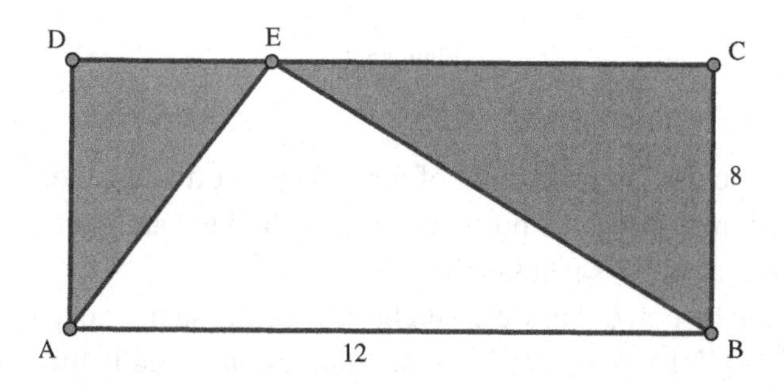

Figure 5.3

A Common Approach

A common approach would be to look at the problem from a different point of view, and instead of finding the area of the shaded region, as being asked for in the problem, we would find the area of the non-shaded region and subtract from the area of the rectangle. The non-shaded triangle, where the base is $AB = 12$", and the altitude is $BC = 8$", has an area of $\frac{1}{2} \cdot 12 \cdot 8 = 48$ square inches. The area of the rectangle is simply $12 \cdot 8 = 96$ square inches. Therefore, the area of the shaded region is simply $96 - 48 = 48$ square inches.

An Exemplary Solution

Another approach using the same strategy is as follows. Since the exact location of point E has not been specified, we can use our problem-solving strategy of considering extreme cases by placing point E so that it coincides with point C as shown in Figure 5.4.

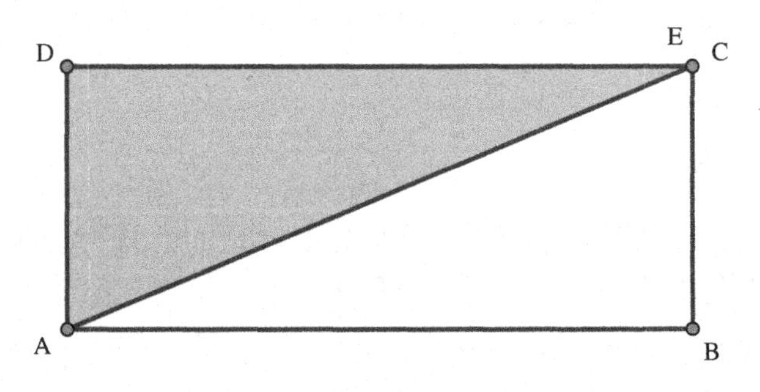

Figure 5.4

Now, since AC is a diagonal of a rectangle, it divides the rectangle in half. Therefore, the shaded portion is exactly half the area of the rectangle, and has an area of 48 square inches.

It should be noted that we could have used the same procedure had the rectangle $ABCD$ been replaced by a parallelogram. Initially that might have made it a bit more challenging, but it could be similarly solved.

PROBLEM 5.7

There are 50 teacher's letterboxes in George Washington High School's general office. One day the letter carrier delivers 151 pieces of mail for the teachers. What is the largest number of letters that any one teacher is *guaranteed* to get?

A Common Approach

It is not uncommon to see an unsuspecting problem solver have a tendency to "fumble around" aimlessly with this sort of problem, usually not knowing where to start. Sometimes, guess and test procedure may work here. But that may not lead to a convincing answer.

An Exemplary Solution

An advisable approach for a problem of this sort is to consider extremes. Naturally, it is possible for one teacher to get all the delivered mail, but this is not *guaranteed.* To best assess this situation we shall consider the extreme case, where the mail is as evenly distributed as possible. This would have each teacher receiving 3 pieces of mail with the exception of one teacher, who would have to receive the 151st piece of mail. Therefore, four pieces of mail is the most any one teacher is *guaranteed* to receive.

PROBLEM 5.8

Point M is the midpoint of side AB of ΔABC. P is any point on AM (Figure 5.5). The line through point M, parallel to PC, meets BC at D. What part of the area of ΔABC is the area of ΔBDP?

A Common Approach

The area of ΔBMC is one-half the area of ΔABC (since the median partitions a triangle into two equal areas). Area of ΔBMC $=$ area $\Delta BMD +$ area $\Delta CMD =$ area $\Delta BMD +$ area ΔMPD, which equals area $\Delta BPD = \frac{1}{2}$ area ΔABC. This rests on the property that when the vertices of two triangles lie on a line parallel to a common base, their areas are equal.

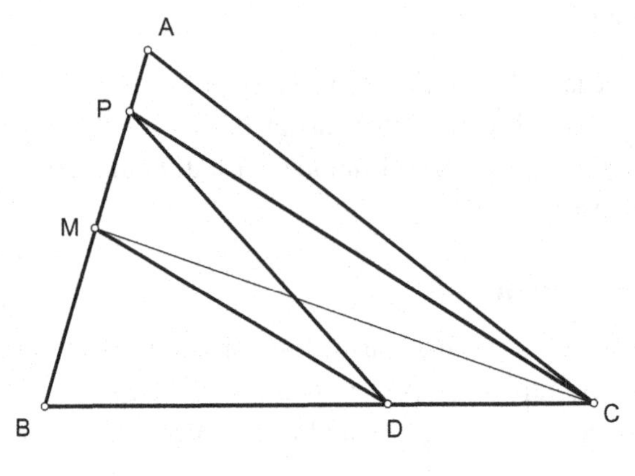

Figure 5.5

An Exemplary Solution

This problem can be considerably simplified by very carefully using our strategy of considering extreme cases. Let us select point P at an extreme position, either at point M or point A. Suppose P is placed at point A. Notice that as P moves along BA towards A, we find that MD which must stay parallel to PC, moves towards a position whereby D approaches the midpoint of BC. That final position for D then has AD as a median of ΔABC. Thus the area of ΔPBD is one-half the area ΔABC, since the median of a triangle divides the triangle into two equal area triangles.

This solution of considering extreme cases provides us with an interesting example where we must watch all movements when we move a point to an extreme position.

PROBLEM 5.9

Two congruent squares, each side of length 4", are placed such that a vertex of one of them is at the center of the other. What is the smallest value for the area of the overlapping section? (Figure 5.6)

Common Approach

The most obvious approach is to make a drawing of the two squares. Some people may actually draw the squares to scale, and attempt to measure the

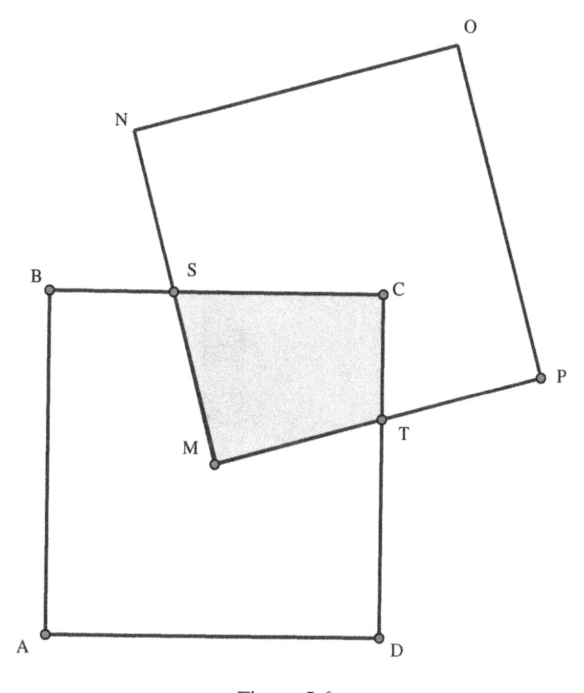

Figure 5.6

results. Since the figure under observation is irregular, measuring the area might prove difficult.

Another common approach is to add some auxiliary lines. One such is to draw line segments *BM* and *CM*. We can easily prove the two triangles *BSM* and *CTM* congruent (*ASA*) (see Figure 5.7). Therefore, quadrilateral *SCTM* is equal in area to triangle *BCM*, since the area of triangle *SCM* is added to the area of the two previously-proved congruent triangles.

An Exemplary Solution

Since the orientation of the squares has not been specified in the problem we can place them anywhere we wish as long as the vertex of one is in the center of the other. Let us use our considering-an-extreme-case strategy. We can place the squares as shown in Figure 5.8, so that the sides of the two squares are mutually perpendicular.

If we are not yet convinced that the shaded region is a quarter of the original square, then all we need to do is to extend line segments *PM* and

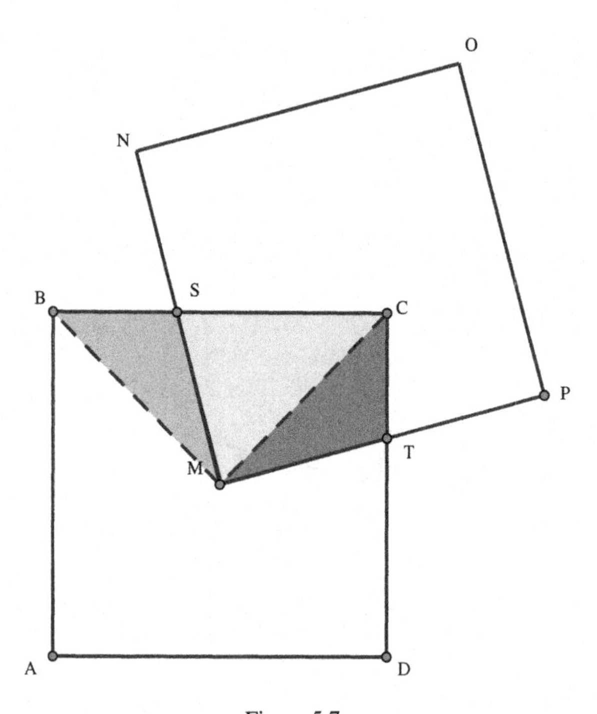

Figure 5.7

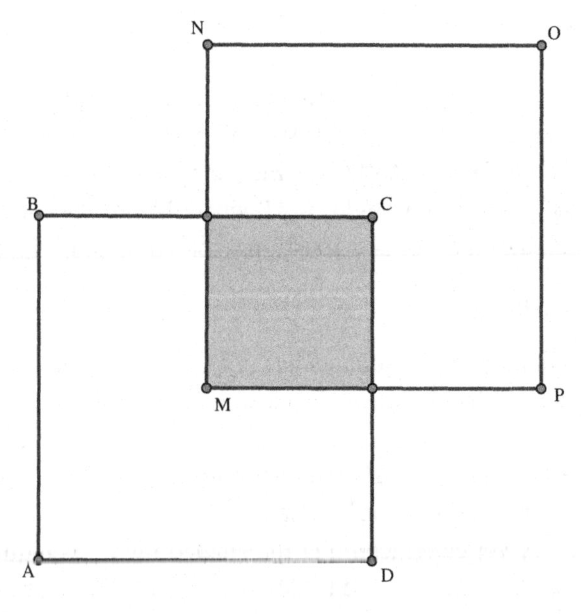

Figure 5.8

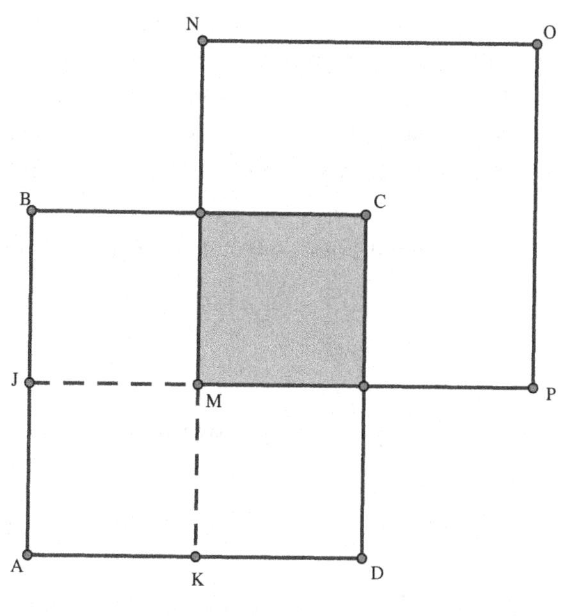

Figure 5.9

NM to meet the sides of the square at points *J* and *K*, respectively as shown in Figure 5.9.

It is apparent that the shaded area is $\frac{1}{4}$ of the original square or $\frac{1}{4}$ of 16 or 4 square inches. By placing the squares in a special position, we were able to easily find the answer to our problem.

PROBLEM 5.10

Find the value of x that satisfies the equation: $x^{x^{x^{x^{x^{\cdot^{\cdot^{\cdot}}}}}}} = 2$.

A Common Approach

At first glance, most people would be overwhelmed, and not know how to approach the problem. This is not a surprising occurrence.

An Exemplary Solution

We could look at this as being somewhat of an extreme situation. We begin by noticing that there is an infinite number of x's in this series, or tower, of powers. Eliminating one of the x's would not have any effect on the end result, because of the nature of infinity. Therefore, by removing the first x, we find that all those remaining in the tower of x's must also equal 2. This then permits us to rewrite this equation as $x^2 = 2$. It then follows that $x = \pm\sqrt{2}$. If we remain in the set of positive numbers, then the answer is $x = \sqrt{2}$.

Below you can see how the successive increases get ever closer to 2.

$$\sqrt{2} = 1.414213562\ldots$$

$$\sqrt{2}^{\sqrt{2}} = 1.632526919\ldots$$

$$\sqrt{2}^{\sqrt{2}^{\sqrt{2}}} = 1.760839555\ldots$$

$$\sqrt{2}^{\sqrt{2}^{\sqrt{2}^{\sqrt{2}}}} = 1.840910869\ldots$$

$$\sqrt{2}^{\sqrt{2}^{\sqrt{2}^{\sqrt{2}^{\sqrt{2}}}}} = 1.892712696\ldots$$

$$\sqrt{2}^{\sqrt{2}^{\sqrt{2}^{\sqrt{2}^{\sqrt{2}^{\sqrt{2}}}}}} = 1.926999701\ldots$$

$$\ldots$$

And so we have a surprisingly simple solution to a very complicated looking problem.

PROBLEM 5.11

Let's Make a Deal was a long-running television game show that featured a problematic situation. A randomly selected audience member would come on stage and be presented with three doors, behind which one had a car and the other two had donkeys. She was asked to select one, hopefully the one behind which there was a car, and not one of the other two doors, each of which had a donkey behind it. If she selected the door with the car, she would be allowed to own the car. There was only one wrinkle in this: After the contestant made her selection, the host, Monty Hall, knowing where the car was located, exposed one of the two donkeys behind a not-selected door (leaving two doors still unopened) and the audience participant was asked if she wanted to stay with her original selection (not yet revealed) or switch to the other unopened door. At this point, to heighten the suspense, the rest of the audience would shout out "stay" or "switch" with seemingly equal frequency. The question is what to do? Does it make a difference? If so, which is the better strategy (i.e., with the greater probability of winning) to use here?

A Common Approach

Some may speculate about what they think intuitively as the best strategy. Most will probably say that there is no difference, since at the end you have a one out of two chance of getting the car. Unfortunately, they are wrong. This will motivate a very curious audience — readers included!

An Exemplary Solution

It might be best to consider this problem step by step, and then to be convincing, we will consider an extreme case to make the crucial point.

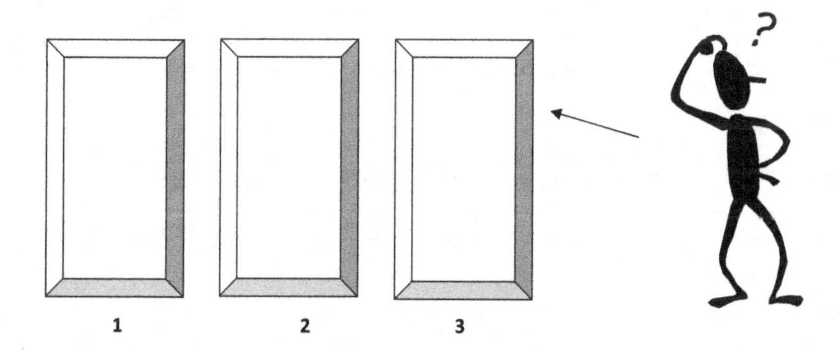

Let us look at this now step-by-step. The result gradually will become clear. There are *two donkeys* and *one car* behind these doors You must try to get the car. You select Door #3 Monty Hall opens one of the doors that you did not select and exposes a donkey.

He asks: "Do you still want your first choice door, or do you want to switch to the other closed door?" To help make a decision, we will use the problem-solving strategy of considering an extreme case. Suppose there were 1000 doors instead of just three doors.

1 2 3 4 • • • 997 998 999 1000

□ □ □ □ □ □ □ □

You choose Door 1000. How likely is it that you chose the right door?

"Very unlikely", since the probability of getting the right door is $\frac{1}{1000}$.

How likely is it that the car is behind one of the other doors (1–999)?

"Very likely": $\frac{999}{1000}$.

1 2 3 4 • • • 996 997 998 999

□ □ □ □ □ □ □ □

These are all "***very likely***" doors!

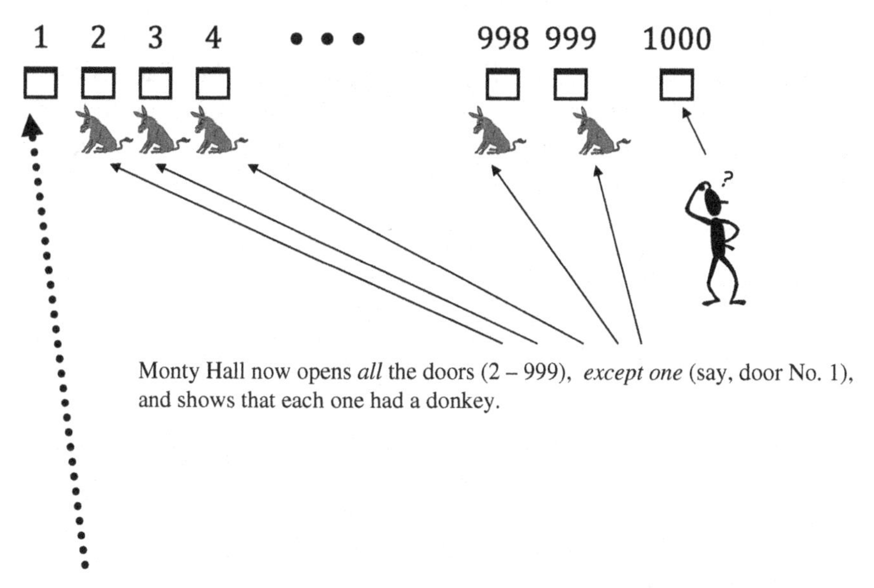

We are now ready to answer the question. Which is a better choice?

- Door 1000 ("*Very unlikely*" door), or
- Door 1 ("*Very likely*" door)?

The answer is now obvious. We ought to select the "very likely" door, which means "switching" is the better strategy for the audience participant to follow. In the extreme case, it is much easier to see the best strategy, than had we tried to analyze the situation with the three doors in the original problem. The principle is the same in either situation.

This problem has caused many an argument in academic circles, and was also a topic of discussion in the New York Times, and other popular publications as well. John Tierney wrote in *The New York Times* (Sunday, July 21, 1991) that "perhaps it was only an illusion, but for a moment here it seemed that an end might be in sight to the debate raging among mathematicians, readers of *Parade* magazine, and fans of the television game show 'Let's Make a Deal'. They began arguing after Marilyn vos Savant published a puzzle in *Parade*. As readers of her 'Ask Marilyn' column are reminded each week, Ms. vos Savant is listed in the Guinness Book of World Records Hall of Fame for 'Highest I.Q.', but that credential did not impress the public when she answered this question from a reader". She gave the right answer, but still many mathematicians argued — yet we solved it!

Chapter 6

Solving a Simpler Analgous Problem

There are problems that appear to be extremely complex when we first approach them. The numbers in the problem might be extremely large, and thus confusing or possibly distracting. Perhaps there is an excessive amount of data given, some of which may not even be necessary to solve the given problem. Even the way the problem has been presented can sometimes confuse the reader. Regardless of the reason, an excellent approach is to convert the problem into a simpler form, but as an equivalent version of the original. Change the numbers, modify the original drawing, or try another approach to simply alter the form of the problem. By solving this simplified version of the problem, the problem solver can possibly gain some insight into how to approach the original problem.

When you first bought a new computer, rather than attack all the features and new capabilities at once, you may have begun with the more familiar features and then gradually added more as you became more acquainted with what the new computer was capable of doing. You started with the simpler features first.

Suppose we were faced with the following problem:

> Given 19 consecutive integers having a sum of 95. What is the 10th integer in the sequence?

Many people would probably use their algebraic skills and write the 19 integers as x, $(x + 1)$, $(x + 2)$, $(x + 3)$, ..., $(x + 17)$, $(x + 18)$ and then add them. Set the result equal to 95 and solve for x. Others may recognize that the 10th integer is the middle number and let it be represented by x. Then the rest of the numbers are represented by $(x + 9)$, $(x + 8)$, $(x+7)$, ..., $(x-7)$, $(x-8)$, $(x-9)$. We can now pair the terms as we add them. That is, pairing $(x - 9)$ and $(x + 9)$ to get $2x$, pairing $(x - 8)$ and

$(x + 8)$, again getting $2x$, and so on, getting $2x$ each time. This is a much simpler version to solve, because you now have the simple equation of $9(2x) + x = 95$, or $19x = 95$, so that $x = 5$.

However, there is an even more interesting approach. Suppose we consider a shorter series of numbers, such as $3 + 4 + 5 + 6 + 7$. Their sum (25) divided by 5 will give you their average, namely, 5, which just happens to be the middle number of the series. For our given series in this posed problem, we find that the 10th term is the middle term, and since the integers are consecutive, it is also the arithmetic mean, or average, of the 19-term sequence. Therefore, to find the average, we simply take the sum, 95, and divide it by the number of terms, 19, to get 5. This simpler version of the problem, allowed us to see the original problem in a much simpler form, and thus making it easier to solve.

Many times, we need not only reduce the complexity of the original problem and solve a simpler version, but we may have to utilize another of our strategies as well. For example, find the decimal value of 1/500,000,000,000.

We cannot use a calculator, since most have a display that does not accept 12 digits. Let us use two of our other strategies: organizing data and finding a pattern. We will solve a series of simpler versions of our problem, and organize the results in a table. Then we can look for a pattern.

Fraction	Number of 0's after the 5	Quotient	Number of 0's between the decimal and 2
$\frac{1}{5}$	0	0.2	0
$\frac{1}{50}$	1	0.02	1
$\frac{1}{500}$	2	0.002	2
$\frac{1}{5000}$	3	0.0002	3
$\frac{1}{50000}$	4	0.00002	4
⋮	⋮	⋮	⋮

There is definitely a pattern evolving here. The number of zeros in the divisor is the same as the number of zeros after the decimal and before the 2. Since there were 11 zeros after the 5 in the divisor, there will be 11 zeros after the decimal point and before the 2:

Fraction	Number of 0's after the 5	Quotient	Number of 0's between the decimal and 2
$\frac{1}{5}$	0	0.2	0
$\frac{1}{50}$	1	0.02	1
$\frac{1}{500}$	2	0.002	2
$\frac{1}{5000}$	3	0.0002	3
$\frac{1}{50000}$	4	0.00002	4
⋮	⋮	⋮	⋮
1/500000000000	**11**	**0.000000000002**	**11**

Notice how the simpler version(s) of the original problem, together with our other two strategies, enabled us to easily solve the given problem. You should be aware that it is not uncommon to make use of more than one strategy to solve a problem.

PROBLEM 6.1

The basketball squad is taking part in a free-throw contest. The first player scored x free throws. The second shooter scored y free throws. The third shooter made the same number of free throws as the arithmetic mean of the number of free throws made by the first two shooters. Each subsequent shooter in the contest scored the arithmetic mean of the number of free throws made by all the shooters who had preceded him. How many free throws did the 12th player make?

A Common Approach

Some may try to solve this problem by finding the arithmetic mean for each of the 12 players in turn. This requires a great deal of time and effort, and it is easy to make an error in the algebraic manipulation. A better solution must be at hand.

An Exemplary Solution

We shall begin by examining a simpler analogous problem. We will replace x and y with simple numbers, and see what happens. Suppose the first player made 8 free throws (x) and the second made 12 free throws (y). Then the third player had a score equal to their arithmetic mean, or $\frac{8+12}{2} = \frac{20}{2} = 10$. Now, the fourth player had a score equal to the arithmetic mean of the first three players, namely, $\frac{8+12+10}{3} = \frac{30}{3} = 10$. Similarly, the score made by the fifth player is the arithmetic mean of the scores of the first four players, $\frac{8+12+10+10}{4} = \frac{40}{4} = 10$. Aha! The score made by any player after the first two will always be the arithmetic mean of the scores of the first two players. The correct answer to the problem is the arithmetic mean of the scores of the first two players, namely, $\frac{x+y}{2}$. The simpler, analogous problem enabled us to determine the method we should use for solving the original problem quite quickly.

PROBLEM 6.2

From any point in or on an equilateral triangle, the sum of the distances to the three sides is the same as for any other such point. What is the sum of these distances, if an equilateral triangle has a side of length 4?

A Common Approach

There are several ways to solve this problem. One of the easiest methods to understand is to select any point in the equilateral triangle (something to be expected by the unsuspecting problem solver), and to draw the three perpendiculars to the sides (Figure 6.1).

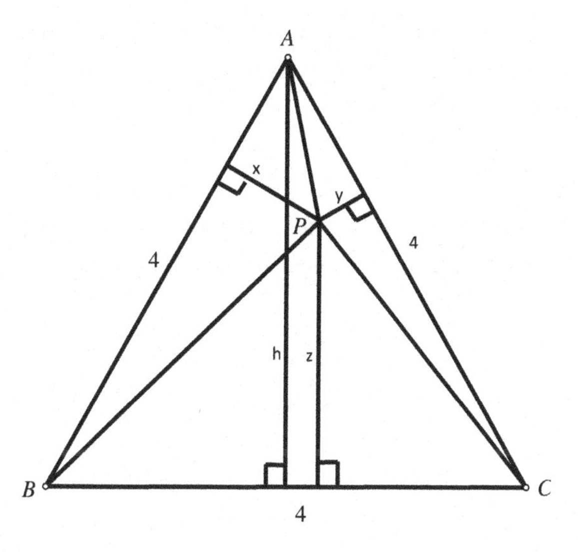

Figure 6.1

By equating the area of ΔABC with the sum of the areas of triangles *APB, PBC*, and *CPA* using the three altitudes x, y, z, and the base 4, we get the area of

$$\Delta ABC = \frac{1}{2}(4)(h) = \frac{1}{2}(4)(x) + \frac{1}{2}(4)(y) + \frac{1}{2}(4)(z) = \frac{1}{2}(4)(x + y + z).$$

Therefore $h = x + y + z$. In this case, we find the altitude of the equilateral triangle is $2\sqrt{3}$. Thus, $x + y + z = 2\sqrt{3}$.

An Exemplary Solution

Without loss of generality, we will consider a simpler analogous problem since we can choose to place the point P anywhere in or on the equilateral triangle as the problem states. Suppose we place P on A, then the solution becomes trivial. The perpendiculars to AB and AC both have length 0, and the perpendicular to BC is simply the altitude of the triangle, or $2\sqrt{3}$.

Note that this solution strategy might also fit our strategy description for considering extreme cases. What we have done is consider the extreme case, where the point assumes the position at the vertex of the triangle. This shows that the selection of a strategy is flexible.

PROBLEM 6.3

In the following expressions, m and n are positive integers, each greater than 1. Which of the following expressions is the greatest?

1. $m+n$
2. $m-n$
3. $\sqrt{2mn}$
4. $\frac{m^2+n^2}{m+n}$
5. $\frac{m^4+n^4}{m^3+n^3}$

A Common Approach

The most obvious approach is to actually perform the operations as given, and attempt to see which one has the greatest value. This is cumbersome and tedious, and requires a great deal of algebraic manipulation.

An Exemplary Solution

Let's solve a simpler version of this problem. We can substitute some convenient positive integral values for the variables, so as to allow us to solve a simpler analogous problem. Suppose we let $m = 2$ and $n = 4$. Then for expression (1) we get $2+4 = 6$. For (2) we get $2-4 = -2$. For (3) we get $\sqrt{16} = 4$. For (4) we get $\frac{4+16}{2+4} = 3.333\bar{3}$. For (5) we get $\frac{16+256}{8+64} = 3.777\bar{7}$. Therefore, we can conclude that the expression of greatest value is $m+n$.

PROBLEM 6.4

We are given that $\frac{1}{x+5} = 4$. What is the value of $\frac{1}{x+6}$?

A Common Approach

The traditional solution is simply to solve the given equation $\frac{1}{x+5} = 4$, and find the value of x, which, it turns out, is $x = -\frac{19}{4}$. We then substitute for x in the fractional expression $\frac{1}{x+6}$, and obtain $\frac{4}{5}$. Of course, this may involve a bit of some cumbersome algebraic and arithmetic manipulations, but it is certainly correct.

An Exemplary Solution

Perhaps a more clever way to approach this problem is to look at the problem from a different point of view, by beginning with the given information: the equation $\frac{1}{x+5} = 4$. If we take the reciprocals of both sides of the equation to obtain $x + 5 = \frac{1}{4}$, we will have something far more manageable. Since we are looking for $x + 6$, we merely have to add 1 to both sides, of this equation, to get $x + 5 + 1 = \frac{1}{4} + 1$, or $x + 6 = \frac{5}{4}$. We again take the reciprocals of both sides to obtain $\frac{1}{x+6} = \frac{4}{5}$, which is what we were asked to find. Clearly this might be considered a more elegant approach.

PROBLEM 6.5

Given a circle and its diameter; show how the area can be partitioned into seven equal area regions without using straight lines.

A Common Approach

Typically when faced with a problem like this, one realizes that the pair of compasses is the tool to use, and begins drawing circles inside the given circle with the hope of somehow seeing a pattern. This will most likely lead nowhere.

An Exemplary Solution

We begin with our given circle and mark off a length along the diameter which is one-seventh of the distance from one endpoint, as shown in Figure 6.2.

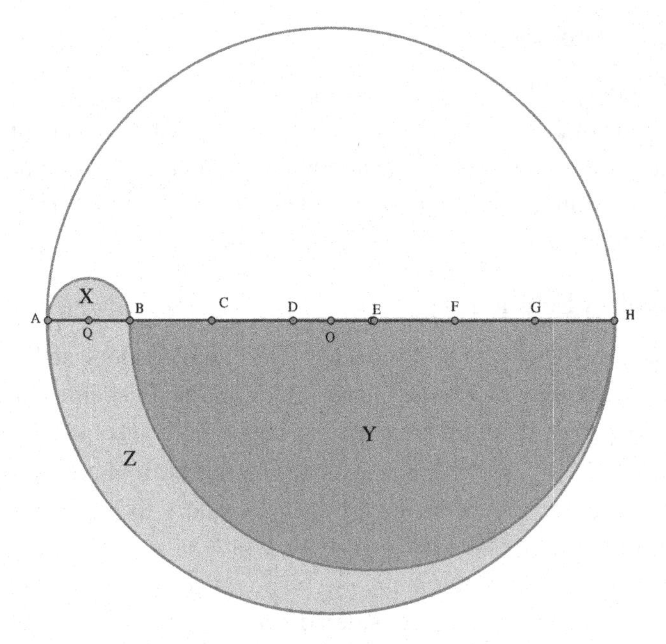

Figure 6.2

We can describe the area of the lighter shaded region as the area of the semicircle of the original circle plus the area of semicircle X minus the area of semicircle Y.

Since we know that the ratio of the areas of circles is related directly to the square of their respective diameters, we can get the following expression to represent the area of the lighter shaded region:

$$\text{Area of } (X+Z) = \text{Area of } (Y+Z) - \text{Area of } (Y).$$

Since the ratio of the diameters of the three semicircles is: $(Y+Z):(Y):(X) = 7:6:1$, their areas are in the ratio of 49:36:1. Using this we can see that the ratio of the lighter shaded region to the larger semicircle is $(49 - 36 + 1):49$ (or 14:49), or written in fractional form the lighter shaded region is $\frac{14}{49}$ of the larger semicircle. In that case the ratio of the lighter shaded region to the whole circle is $\frac{1}{2} \cdot \left(\frac{14}{49}\right) = \frac{7}{49} = \frac{1}{7}$. We multiply by $\frac{1}{2}$ since the ratio of $\frac{14}{49}$ is the ratio of $\frac{1}{2}$ of the full circle.

Using this strategy, we can then consider the semicircles with diameters: AC, AD, AE, AF, AG, and AH, which, taken together, will partition the circle into sevenths.

PROBLEM 6.6

Two trains, one going from Chicago to New York and the other from New York to Chicago, a distance of 800 miles, one traveling uniformly at 60 miles per hour, and the other at 40 miles per hour, start towards each other, at the same time, along the same track. At the same time a bee begins to fly from the front of one of the trains, at a speed of 80 miles per hour towards the on-coming train. After touching the front of this second train, the bee reverses direction and flies towards the first train (still at the same speed of 80 miles per hour). The bee continues this back and forth flying until the two trains collide, crushing the bee. How many miles did the bee fly?

A Common Approach

This problem might remind the reader of the kind of verbal problem found in most algebra textbooks; however, there is a curious twist to this that would not be found in one of those typical uniform-motion problems. One is naturally drawn to find the individual distances that the bee traveled. An immediate reaction is to set up an equation based on the familiar relationship: "rate times time equals distance". However, this back and forth path is rather difficult to determine, requiring considerable calculation. Even then, it is very difficult to solve the problem in this fashion.

An Exemplary Solution

A much more elegant approach would be to solve a simpler analogous problem (one might also say, we can look at the problem from a different point of view). We seek to find the distance the bee traveled. If we knew the time the bee traveled, we could determine the bee's distance because we already know the speed at which the bee is flying.

The time the bee traveled can be easily calculated, since it traveled the entire time the two trains were traveling — until they collided. To determine the time, t, the trains traveled, we set up an equation as follows.

The distance of the first train is $60t$, and the distance of the second train is $40t$. The total distance the two trains traveled is 800 miles. Therefore, $60t + 40t = 800$, and $t = 8$ hours, which is also the time the bee traveled. We can now find the distance the bee traveled, which is $8 \cdot 80 = 640$ miles. What seemed to be an incredibly difficult task, that of finding the distance the bee traveled back and forth, has been reduced to a rather simple application of a common "uniform-motion problem" of the sort encountered in the elementary algebra course with the solution readily apparent.

PROBLEM 6.7

Given a randomly-drawn pentagram as shown in Figure 6.3, what is the sum of the acute angles at each of the vertices of the pentagram.

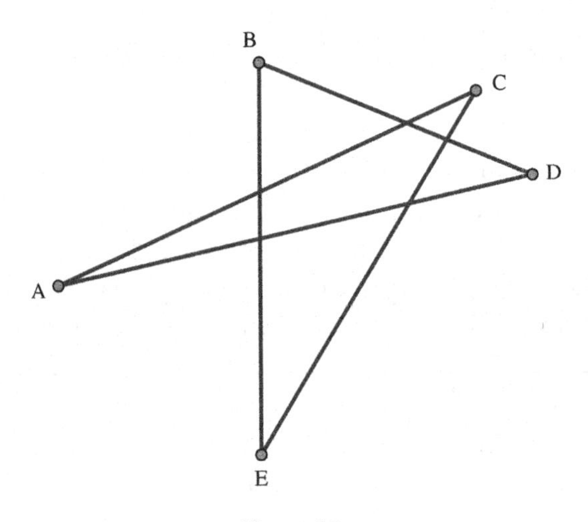

Figure 6.3

A Common Approach

Unfortunately, most folks in order to attack this problem would search for a protractor to measure each of the angles — hopefully accurately — and then take this sum. From this they would try to make an intelligent guess as to what that sum would be.

An Exemplary Solution

We could use our strategy of solving a simpler analogous problem. That is, since this pentagram was not specified as to shape or regularity, we could assume that it is a pentagram that is inscribed in a circle, as shown in Figure 6.4. If we inspect each of the acute angles of the pentagram we will notice that they are each an inscribed angle of the circle, which means that each one's measure is one-half the measure of its intercepted arc. For example, $\angle A = \frac{1}{2} arcCD$. When we look at the arcs of each of the remaining four acute angles of the pentagram, we find that we then have the sum of the arcs comprising the entire circle. The sum of the angle measures is one-half the sum of the measures of each of the arcs comprising the circle, which is essentially the same as one-half of the measure of the circle, or 180°.

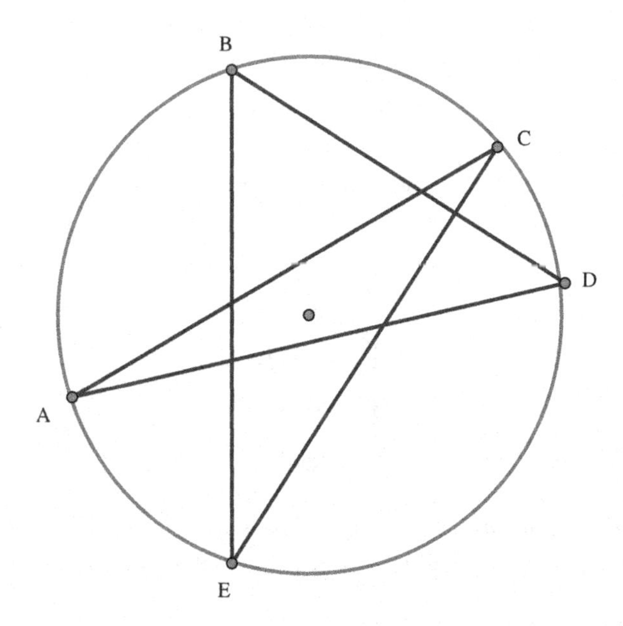

Figure 6.4

PROBLEM 6.8

Which of the following has the largest value?

$$1^{48}, 2^{42}, 3^{36}, 4^{30}, 5^{24}, 6^{18}, 7^{12}, 8^{6}$$

Common Approach

With a computer program or even a calculator, which can read out many digits, we can attempt to actually compute the value of each expression. However, this will prove to be a very long and tedious approach. Yet it can be done this way.

An Exemplary Solution

Let us use our strategy for solving a simpler analogous problem. A quick inspection of the numbers given reveals that each of the exponents is a multiple of 6. If we take the 6th root of each term (or raise each to the $\frac{1}{6}$ power), we can simplify the terms to be compared. That is, we know that the original numbers given are all of the following to the 6th power. Therefore, the largest of the following will be associated with the largest of the original numbers we were asked to compare.

$$1^{8}, 2^{7}, 3^{6}, 4^{5}, 5^{4}, 6^{3}, 7^{2}, 8^{1}.$$

The values of these can be relatively easy to calculate:

$$2^7 = 128,$$

$$3^6 = 729,$$

$$4^5 = 1024,$$

$$5^4 = 625,$$

$$6^3 = 216.$$

The remaining numbers are clearly smaller. Thus, the largest of the original set of 8 power expressions is 4^{30}, which came from $(4^5)^6$.

PROBLEM 6.9

To extend the amount of wine in a 16-ounce bottle, David decides upon the following procedure. On the first day, he will only drink 1 ounce of the wine, and then refill the bottle with water. On the second day, he will drink 2 ounces of the water–wine mixture, and then again refill the bottle with water. On the third day, he will drink 3 ounces of the water–wine mixture, and again refill the bottle with water. He will continue this procedure for succeeding days until he empties the bottle by drinking 16 ounces of mixture on the 16th day. How many ounces of water will David drink altogether?

A Common Approach

It is very easy to get bogged down with a problem like this one. Some readers may begin to make a table showing the amount of wine and water in the bottle on each day, and attempt to compute the proportional amounts of each type of liquid David drinks on any given day. We could more easily resolve the problem by examining it from another point of view, namely, how much water does David add to the mixture each day? Since he eventually empties the bottle (on the 16th day), and it held no water to begin with, he must have consumed all the water that was put into the bottle. On the first day, David added 1 ounce of water. On the second day, he added 2 ounces of water. On the third day, he added 3 ounces of water. On the 15th day, he added 15 ounces of water. (Remember, no water was added on the 16th day.) Therefore, the number of ounces of water David consumed was $1 + 2 + 3 + 4 + 5 + 6 + 7 + 8 + 9 + 10 + 11 + 12 + 13 + 14 + 15 = 120$ ounces.

An Exemplary Solution

While this solution is indeed valid, a slightly simpler analogous problem to consider would be to find out how much liquid David drank altogether, and then simply deduct the amount of wine, namely 16 ounces.

Thus, $1 + 2 + 3 + 4 + 5 + 6 + 7 + 8 + 9 + 10 + 11 + 12 + 13 + 14 + 15 + 16 = 136$, and $136 - 16 = 120$.

David consumed 136 ounces of liquid, of which 120 ounces was the water.

Chapter 7

Organizing Data

The strategy of organizing data is one of the most important strategies, even though at first glance we might assume it to be axiomatic. That is, everyone should automatically organize the data that appears in a problem without giving it a moment's thought. It is what we often do in real life without consciously thinking about it.

For example, when we begin to prepare our tax forms every spring, we automatically organize our data without any prompting. The way in which we organize our receipts, checks, W-2 forms and so on, makes quite a difference in filling out the complex tax forms.

Most people prepare a carefully organized shopping list before leaving for the market. They might organize their needs by categories, by location in the market, or by necessity. Similarly, when on a vacation trip, we are most likely to make an organized list of what we want to see. In both cases, the list might be written or might simply be in your mind.

When the major polling organizations collect data from a poll, it is not at all unusual for two or more polls to yield different results, depending on how the same data were organized by each organization.

When people are faced with problems containing a lot of data, they often become confused by the way in which the data are presented within the problem. Learning to organize data in a meaningful and clear manner is a problem-solving tool of major importance. Let us look at a problem that utilizes this strategy.

> A group of archaeologists was digging at a site. Here are the number of pieces of pottery they dug up each day for 15 consecutive days:
>
> $$13, 45, 12, 47, 8, 18, 13, 27, 98, 11, 23, 67, 51, 14, 6.$$
>
> What is the median number of pottery pieces they dug up?

It would probably be next to impossible to resolve this problem the way the number of items found each day were written down. However, let us organize the data in a more meaningful way — from the lowest to the highest:

$$6, 8, 11, 12, 13, 13, 14, 18, 23, 27, 45, 47, 51, 69, 98.$$

Now it is easy to find the median. It is the middle score, or in this case, the eighth score, or 18.

Let us consider another problem, where organizing the data is valuable.

> Jack and Marlene each want to join a DVD movie club. They have two offers. The Freedom Movie Club charges a membership initiation fee to join the club of $20. They then charge $6.20 per movie. The New Look Club has no initiation fee, but charges $8.10 per DVD. Jack decides to join the Freedom Club while Marlene joins the New Look Club. How many DVDs must they each buy, before Marlene has spent more than Jack? How much more will she have spent?

To resolve the problem, we can organize the data with a trio of lists:

Number of DVDs	Freedom Club	New Look Club
0	$20.00	$0.00
1	$26.20	$8.10
2	$32.40	$16.20
3	$38.60	$24.30
4	$44.80	$32.40
5	$51.00	$40.50
6	$57.20	$48.60
7	$63.40	$56.70
8	$69.60	$64.80
9	$75.80	$72.90
10	$82.00	$81.00
11	**$88.20**	**$89.10**
12	$94.40	$97.20

When they each buy the 11th DVD, Marlene will have spent more than Jack. Marlene will have spent $89.10–$88.20 or 90¢ more. It is easy to obtain the answer to both parts of the problem by examining the organized data in the table.

Here is a geometry problem that can only be solved by carefully organizing the data.

> A triangle has integral sides and a perimeter of 12. What are the lengths of its three sides?

Let us prepare an organized list using A, B, and C to represent the sides of the triangle. We will start with A = 1 and list all possibilities for A = 1. Then we will move on to A = 2 and so on.

A	1	1	1	1	1	2	2	2	2	3	3	4
B	1	2	3	4	5	2	3	4	5	3	4	4
C	10	9	8	7	6	8	7	6	5	6	5	4

This list contains all of the number triples whose sum equals 12. But, remember, in a triangle the sum of any two sides must always be greater than the third side or the triangle cannot exist. This eliminates most of the choices. The only three possibilities are 2–5–5, 4–4–4, and 3–4–5. Making an organized list made solving the problem easy.

In this chapter we will present problems that can be most efficiently solved by organizing the data in some meaningful manner. Although some of these problems can be solved with other methods, we merely offer these as a way to expose the advantages to this seemingly unorthodox method of solution.

PROBLEM 7.1

There is a free-throw shooting contest between halves of the basketball tournament. The two finalists are Robbie and Sandy. The first one to sink two consecutive free throws or a total of three free throws altogether will be the winner. In how many different ways can a winner be declared?

A Common Approach

Most people will begin by attempting to find all possible combinations that would result in winning. But, how will we know if we have listed them all? It sounds like a very cumbersome task.

An Exemplary Solution

Let us use our strategy of organizing the data, and make two exhaustive lists of the ways each player can win. The first list shows the ways Robbie can make the first free throw; the second list shows the ways if Sandy makes the first free throw.

R R	S S
R S S	S R R
R S R R	S R S S
R S R S R	S R S R S
R S R S S	S R S R R

There are 10 possible ways the free-throw contest can end. The exhaustive list revealed all the possible ways in a neat, orderly manner.

PROBLEM 7.2

How many triangles are in Figure 7.1?

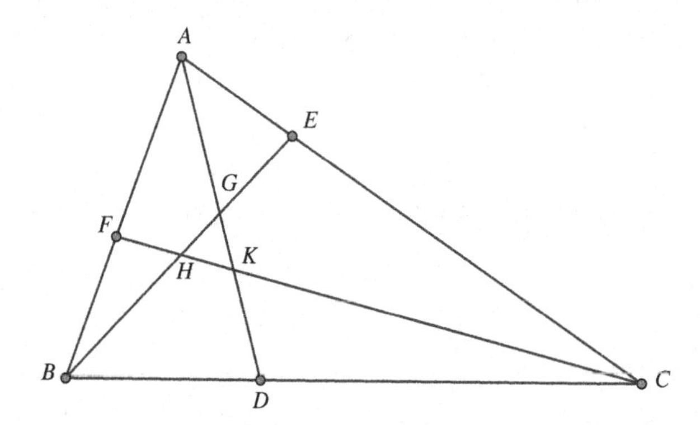

Figure 7.1

A Common Approach

Typically one would start counting the various triangles in some orderly fashion, but not with a specific organized arrangement. Too often, this leads to confusion and uncertainty as to whether all of the triangles have been counted. Then there is the traditional method, or that which involves formal counting methods. These involve calculating the combinations that can be formed by the six lines and excluding those combinations that result in concurrency. Hence, the number of combinations of six lines taken three at a time yield ${}_6C_3 = 20$. From this we subtract the three concurrencies (at the vertices). Thus, there are 17 triangles in the figure.

An Exemplary Solution

Let us try to simplify the problem by reconstructing the figure, gradually adding the lines as we go, and counting from this form of organized data. That is, counting the triangles created by the addition of each additional part of the figure. We can start with the original triangle, *ABC*. Thus, we have exactly one triangle.

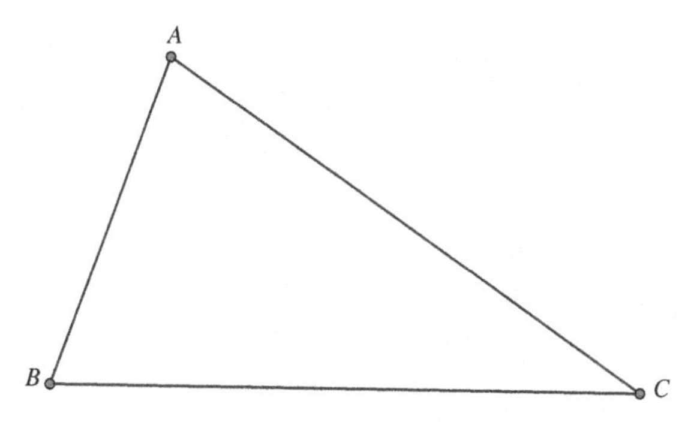

Figure 7.2

Now we will consider the triangle *ABC* with one interior line segment, *AD*. We now have two new triangles, *ABD* and *ADC*.

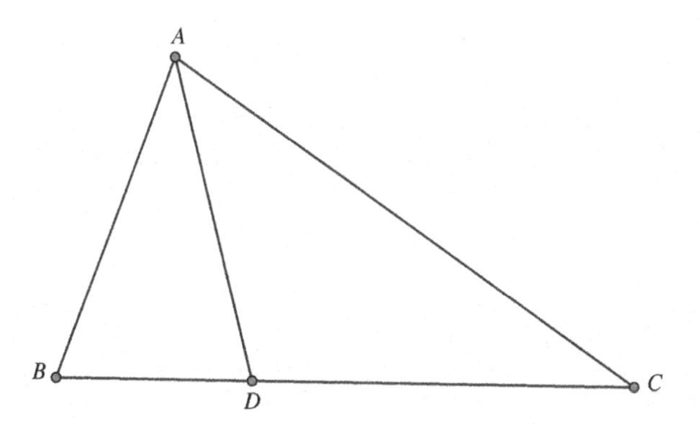

Figure 7.3

Now we can add the next interior line segment, BE, and count all the new triangles that use BE as a side.

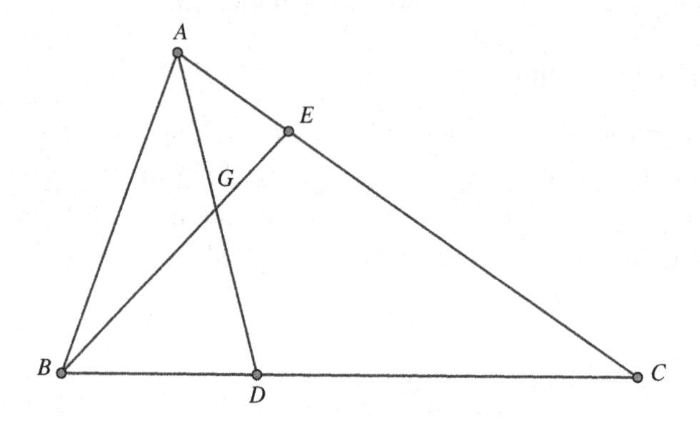

Figure 7.4

Continue in this manner, we now add line segment CF. Again, we will count the new triangles that use part of CF as a side.

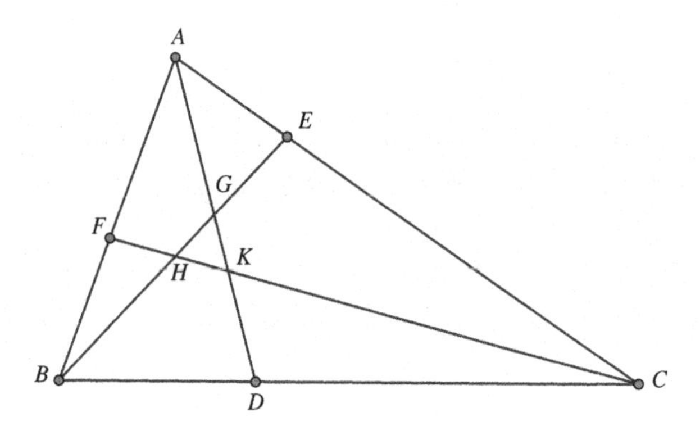

Figure 7.5

Let us put these results into a table.

Figure	Added Line Segment	New Triangles Formed
7.2	0	*ABC*
7.3	*AD*	*ABD*, *ADC*
7.4	*BGE*	*ABG*, *BGD*, *AGE*, *BEC*, *ABE*
7.5	*CKHF*	*FBH*, *AFC*, *BHC*, *AFK*, *KDC*, *AKC*, *FBC*, *HKG*, *EHC*

The total number of triangles listed above is 17.

PROBLEM 7.3

We are given the following sequence: $10^{\frac{1}{11}}, 10^{\frac{2}{11}}, 10^{\frac{3}{11}}, 10^{\frac{4}{11}}, \ldots, 10^{\frac{n}{11}}$ and the problem is to find the smallest positive integer n so that the product of the first n terms of this sequence is greater than 100,000.

A Common Approach

More than likely, the common approach is to use a trial and error approach and consistently take additional members of this sequence and multiply them, each time adding another successive member, until the number 100,000 is finally exceeded. This is obviously a laborious task, and by no means an elegant solution.

An Exemplary Solution

We will begin by taking the product of the first n terms of the given sequence, which, in a fashion, is organizing our data into a manageable form:

$$10^{\frac{1}{11}} \cdot 10^{\frac{2}{11}} \cdot 10^{\frac{3}{11}} \cdot 10^{\frac{4}{11}} \cdot \ldots \cdot 10^{\frac{n}{11}} = 10^{\frac{(1+2+3+4+\cdots+n)}{11}} = 10^{\frac{n(n+1)}{22}}.$$

We realize that "exceed 100,000" means that we will have to exceed10^5, and this happens only when $\frac{n(n+1)}{22} > 5$, or $n(n+1) > 110$. When $n \leq 10$ we have $n(n+1) \leq 110$. Therefore, the smallest integer that n can take on to meet our requirements is 11.

PROBLEM 7.4

Jerome has just opened a kayak franchise. He rents kayaks by the hour. He has to paint identification numbers on each kayak. Each kayak will have three numerals. The first numeral must be his franchise numeral, namely 1. No digits can be repeated on any one kayak, and the three numerals must be in ascending order. No zeros are used. He found that he had painted all the possible combinations that satisfy the requirements. What is the maximum number of kayaks Jerome can have?

A Common Approach

The most usual approach is to begin writing all possible 3-digit numbers that fit the given conditions. But how will we know when we have them all? Is there an order to follow to assure us? The common approach is clearly not very efficient!

An Exemplary Solution

Let us use a carefully made list to organize our data:

First Digit	Second Digit	Third Digit	Number of Choices
1	2	(3 through 9)	7
1	3	(4 through 9)	6
1	4	(5 through 9)	5
1	5	(6 through 9)	4
1	6	(7 through 9)	3
1	7	(8 through 9)	2
1	8	(9)	1

He will have no more than $7 + 6 + 5 + 4 + 3 + 2 + 1 = 28$ kayaks.

PROBLEM 7.5

A farmer is bringing boxes of apples from his farm to market. He has six boxes of apples. However, the scale at the weighing station only weighs five boxes at a time. We are given the results of six separate weighings:

$$\begin{aligned}
&\text{box B} + \text{box C} + \text{box D} + \text{box E} + \text{box F} = 200 \text{ pounds},\\
&\text{box A} + \text{box C} + \text{box D} + \text{box E} + \text{box F} = 220 \text{ pounds},\\
&\text{box A} + \text{box B} + \text{box D} + \text{box E} + \text{box F} = 240 \text{ pounds},\\
&\text{box A} + \text{box B} + \text{box C} + \text{box E} + \text{box F} = 260 \text{ pounds},\\
&\text{box A} + \text{box B} + \text{box C} + \text{box D} + \text{box F} = 280 \text{ pounds},\\
&\text{box A} + \text{box B} + \text{box C} + \text{box D} + \text{box E} = 300 \text{ pounds}.
\end{aligned}$$

How many pounds of apples are in each of the six boxes?

A Common Approach

This problem could be solved algebraically, by setting up a series of six equations in six variables as show here:

$$\begin{aligned}
B + C + D + E + F &= 200,\\
A + C + D + E + F &= 220,\\
A + B + D + E + F &= 240,\\
A + B + C + E + F &= 260,\\
A + B + C + D + F &= 280,\\
A + B + C + D + E &= 300.
\end{aligned}$$

Solving these six equations simultaneously requires a considerable amount of tedious work, so there must be a better way to approach this problem.

An Exemplary Solution

We can make use of our strategy of organizing data in a way that will make the solution relatively simple, and thereby elegant. We begin by organizing the data from the problem in a table such as the one that follows:

Weighing	Box A	Box B	Box C	Box D	Box E	Box F	Total
1	—	B	C	D	E	F	200
2	A	—	C	D	E	F	220
3	A	B	—	D	E	F	240
4	A	B	C	—	E	F	260
5	A	B	C	D	—	F	280
6	A	B	C	D	E	—	300

As we look at this seemingly unmanageable list of equations, we can consider this from a different point of view by organizing the data vertically — taking the vertical sums:

$$5A + 5B + 5C + 5D + 5E + 5F = 1500.$$

When we divide both sides of the equation by 5 we get

$$A + B + C + D + E + F = 300.$$

But the sixth weighing on the table shows that $A + B + C + D + E = 300$ pounds. Therefore, box F must weigh 0 pounds. We then notice that weighing 5 shows that $A + B + C + D + F = 280$ pounds, but since $\mathbf{F = 0}$, we can conclude that $A + B + C + D = 280$.

Recall that the sixth weighing tells us that $A + B + C + D + E = 300$. If we subtract these last two equations, we find that $\mathbf{E = 20}$ pounds.

Recall that weighing 4 gave us: $A + B + C + E + F = 260$, and substituting the previously determined values of F and E, we have $A + B + C + 20 + 0 = 260$, or $A + B + C = 240$. By substituting this value for $A + B + C$ into weighing 5, we find that $\mathbf{D = 40}$.

If we subtract the equations of weighing 3 from weighing 4, and knowing that $F = 0$, we get:

$$\begin{array}{r} A + B + D + E + F = 240 \\ A + B + C + E + F = 260 \\ \hline C - D = 20 \end{array}$$

Since $D = 40$, we have $\mathbf{C = 60}$.

Using weighing 1: $B + C + D + E + F = 200 = B + 60 + 40 + 20 + 0$, so that $\mathbf{B = 80}$.

Similarly, using weighing 2, we get $\mathbf{A = 100}$.

Organizing of the data into a table, made the data manageable, so we could logically resolve the problem.

PROBLEM 7.6

Consider 3-digit numbers where all the digits are odd numbers. What is the sum of all these numbers?

A Common Approach

Typically when faced with a problem like this, one is apt to begin to list many of these odd numbers in some organized fashion, then revert to doing this tedious addition.

An Exemplary Solution

The key to a clever method of solution is to organize these numbers in some manageable fashion. If we were to look at the list — writing them in some well-organized way — our list could look like the following: $111 + 113 + 115 + 117 + 119 + 133 + 135 + 137 + 139 + \cdots + 511 + 513 + 515 + 517 + 519 + \cdots + 991 + 993 + 995 + 997 + 999$. Since there are five digits that can occupy each of the three places, there are $5 \cdot 5 \cdot 5 = 125$ possible numbers. If we look at these in an organized fashion, we can add them in pairs: the first and the last, the second and the next-to-last, etc. The sum of each of these pairs is 1110. There are $\frac{125}{2}$ pairs in this list. Therefore, the sum of these numbers is $\frac{125}{2} \times 1110 = 69{,}375$.

One can also organize this data in another way that would also lead to a relatively elegant solution. We have established that there are 125 such integers that need to be added, each of which has three digits resulting in a total of 375 digits to be considered. Clearly each of the five odd digits: 1, 3, 5, 7, and 9 appears 75 times — i.e., 25 times in each of the hundreds, tens, and units places. Symbolically, we can write this as

$$25[100(1+3+5+7+9) + 10(1+3+5+7+9) \\ +1(1+3+5+7+9)] = 25 \cdot 25 \cdot (100 + 10 + 1) = 69{,}375.$$

In each of the above cases organizing the data has made the solution of the problem significantly more elegant than just using “brute force” to solve the problem.

PROBLEM 7.7

Suppose we have 11 lines drawn in a plane, where exactly three lines pass through the point P, and exactly three lines contain point Q. Of all these 11 lines, no other three lines are concurrent. What is the minimum number of intersection points that these 11 lines will make under these conditions?

A Common Approach

The most common way of approaching this problem is by trial and error and with 11 lines this can become somewhat confusing. Consequently, there must be a more efficient way of approaching this problem.

An Exemplary Solution

To most efficiently solve this problem we will have to organize the lines in a logical fashion. We would begin by first drawing the three lines that will meet at point P, as shown in Figure 7.6.

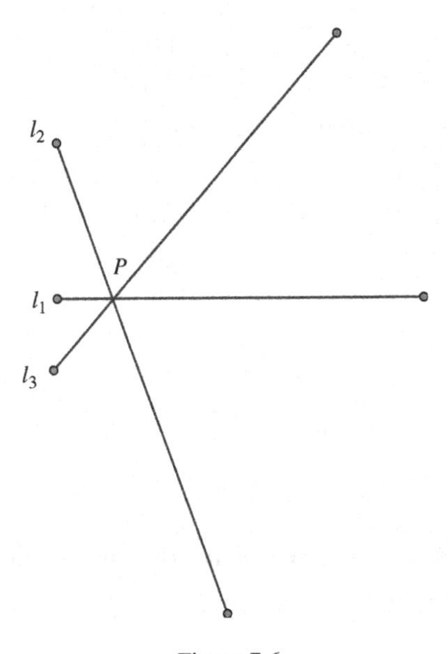

Figure 7.6

We now repeat this process with point Q, as shown in Figure 7.7, by drawing lines $l_3 \| l_4$ and $l_2 \| l_5$.

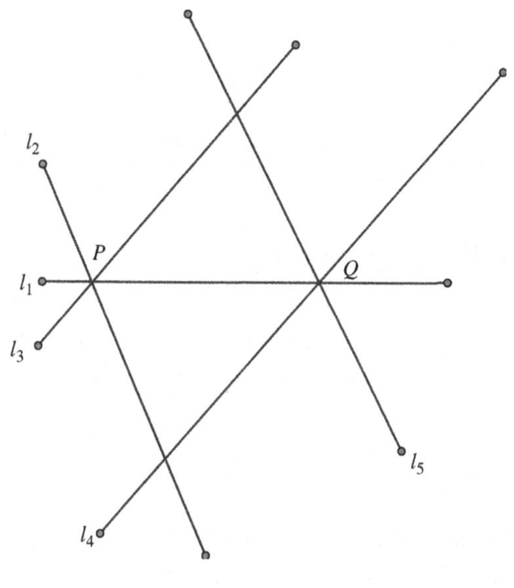

Figure 7.7

We then insert the remaining six lines, making them parallel to line l_2. We show this in Figure 7.8. Each of these lines creates an additional three points of intersection.

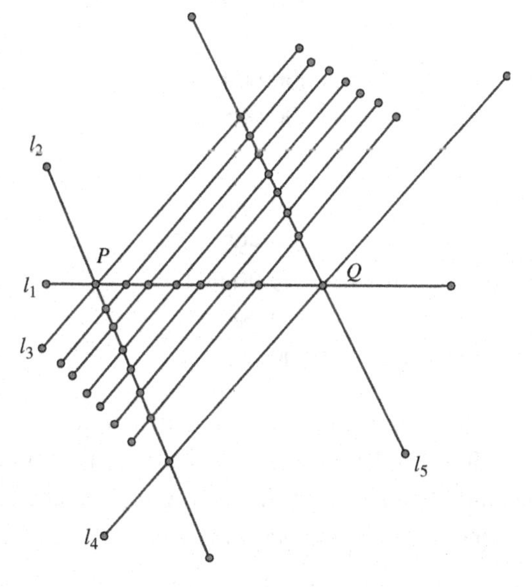

Figure 7.8

Therefore, now having organized our given data in a meaningful fashion, the number of points of intersection is equal to $6(3) + 4 = 22$.

PROBLEM 7.8

When all the numbers 1 to 1,000,000 are printed, how many times will the digit 8 be printed in this list of numbers?

A Common Approach

A typical response to this question, which appears to be overwhelming in nature, is to begin to list all the numbers without looking for any form of organization. Reaching an answer here might depend on stumbling on some good logic.

An Exemplary Solution

The best strategy here would probably be to use that of organizing our data in the following fashion so that we can determine any consistencies in the list

000 000
000 001
000 002
000 003
000 004
000 005
⋮
999 996
999 997
999 998
999 999
000 000

There are 6 million digits listed above, with a digits $0, 1, 2, 3, 4, \ldots, 8$, and 9 are used an equal number of times. We can see this since each "pattern" of digits can be made from each of the ten digits. Therefore, the number 8 will appear $\frac{1}{10}$ of the time, or 600,000 times.

PROBLEM 7.9

A clock-maker has two clocks that strike 12 o'clock midnight at the exact same time. However, one of the clocks gains 1 minute every hour, while the other loses 1 minute every hour. If they continue at this rate, at what time will they read the same?

A Common Approach

The typical immediate approach is to attempt to write an equation. If we let x represent the time until the clocks read the same, we obtain $12 + x = 12 - x$. Solving, we obtain $2x = 0$ and $x = 0$. Not much help.

An Exemplary Solution

In each 24-hour day, clock A will gain 24 minutes while clock B will lose 24 minutes. Thus in five days, clock A will gain $5 \times 24 = 120$ minutes, or 2 hours. Meanwhile, clock B will lose 120 minutes, or 2 hours, every five days. Let us use our strategy of organizing data with the following table to keep account of the hour by hour times.

Day	1	2	5	10	15
Time	$1 \cdot 24 = 24$ mins.	$2 \cdot 24 = 48$ mins.	$5 \cdot 24 = 2$ hrs.	$10 \cdot 24 = 4$ hrs.	$15 \cdot 24 = 6$ hrs.
Gains	12:24 a.m.	12:48 a.m.	2:00 a.m.	4:00 a.m.	6:00 a.m.
Losses	11:36 p.m.	11:12 p.m.	10:00 p.m.	8:00 p.m.	6:00 p.m.

As the table shows, both will read 6:00 at the end of the 15th day. Of course, one clock will be showing 6:00 a.m., while the other reads 6:00 p.m. Nevertheless, they both show 6:00, which is what the problem asked for.

PROBLEM 7.10

How many positive 3-digit odd integers are there such that the product of the three digits of the number is 252?

A Common Approach

The most common approach is to find the factor triples of 252. That is, all sets of three numbers whose product is 252. We should do this in an organized manner, beginning with 1, 1, 252 then 1, 2, 126, then 1, 3, 84 then 1, 4, 63 and so on. We will keep on in this manner, until we find at least one set that gives us a 3-digit odd number. But, probably there is more than one such set. How can we tell when we have them all? This "brute force" method is really not very efficient.

An Exemplary Solution

Let us approach this by using our strategy of organizing the data. We can factor 252 into $2 \times 2 \times 3 \times 3 \times 7$. If one of the digits is 7, then the other digits must have a product of 36 — that is 4 and 9, or 6 and 6 — so as to take into account the rest of the factors, since any other combination of these factors would result in numbers of more than one digit. Combining these digits together with the 7, we find there are five numbers meeting the requirements. These are 749, 479, 947, 497, 667. All of these are 3-digit odd numbers, and the product of their digits is 252 as required.

PROBLEM 7.11

Of the following, which is the largest and the second largest value?

$$\sqrt{2},\ \sqrt[3]{3},\ \sqrt[8]{8},\ \sqrt[9]{9}.$$

A Common Approach

In today's world one goes immediately for a calculator to determine the answer to this problem. However, this may not be so easy with many calculators since the root extraction may not be possible.

An Exemplary Solution

To begin it may be more manageable to change each of these four terms to be shown with fractional exponents as follows:

$$2^{\frac{1}{2}},\ 3^{\frac{1}{3}},\ 8^{\frac{1}{8}},\ 9^{\frac{1}{9}}.$$

The problem-solving strategy that works best here is to organize data in such a way that we can more easily compare expressions through similar exponents.

Since $(2^{\frac{1}{2}})^8 = 2^4 = 16$, and $(8^{\frac{1}{8}})^8 = 8$, we have $(2^{\frac{1}{2}})^8 > (8^{\frac{1}{8}})^8$, or $(2^{\frac{1}{2}}) > (8^{\frac{1}{8}})$. Since $(2^{\frac{1}{2}})^{18} = 2^9 = 512$, and $(9^{\frac{1}{9}})^{18} = 9^2 = 81$, we have $(2^{\frac{1}{2}})^{18} > (9^{\frac{1}{9}})^{18}$, or $(2^{\frac{1}{2}}) > (9^{\frac{1}{9}})$. Since $(3^{\frac{1}{3}})^6 = 3^2 = 9$, and

$(2^{\frac{1}{2}})^6 = 2^3 = 8$, we have $(3^{\frac{1}{3}})^6 > (2^{\frac{1}{2}})^6$, or $(3^{\frac{1}{3}}) > (2^{\frac{1}{2}})$. We can, therefore, conclude that because $(3^{\frac{1}{3}}) > (2^{\frac{1}{2}})$, and $(2^{\frac{1}{2}})$ is greater than both $(8^{\frac{1}{8}})$ and $(9^{\frac{1}{9}})$, we can further conclude that of these four terms the largest is $(3^{\frac{1}{3}})$ and the next larger is $(2^{\frac{1}{2}})$.

PROBLEM 7.12

The class is going on a field trip. There are five children who want to go, but only three openings. The five children are Amanda, Bill, Carol, Dan and Evan. Their teacher put five slips of paper, each with one of their names on it, in a hat and pulled out three at random. What is the probability that Amanda, Bill and Carol will be chosen to go on the trip?

A Common Approach

First, let us find out how many different ways the three people can be chosen. Order does not matter, so this is a combinations problem. Five slips chosen 3 at a time:

$$_5C_3 = \frac{5 \cdot 4 \cdot 3}{1 \cdot 2 \cdot 3} = 10.$$

Since only one of these ways would be Amanda, Bill and Carol, the answer is $\frac{1}{10}$.

An Exemplary Solution

If you do not remember the way to do combinatorials, we can use our organizing-data strategy. We will make a list of all possible ways to choose three names, ignoring order:

ABC	BCD	CDE
ABD	BCE	
ABE	BDE	
ACD		
ACE		
ADE		

There are ten possible ways to select three people. Only one of them, namely, ABC, satisfies the given condition. Thus the correct answer is one of ten, or $\frac{1}{10}$.

Chapter 8

Making a Drawing or Visual Representation

When a problem asks questions about a particular given geometric figure or drawing, it goes without saying that the drawing or visual representation is an integral part of the solution method. It is necessary, and helps to solve the problem. In that sense it is hard to imagine that some mathematicians in ancient times often derived geometric concepts without drawings — or at least they presented their geometric findings without the appropriate sketch. However, there are many problems where a drawing is not implicit in the statement of the problem; yet, actually *seeing* what is being considered can be of great help. Many people are visual learners; they need a picture rather than merely words to understand what is taking place. And no, it is not a case of daydreaming. Contrary to what some people believe, visualization has nothing to do with daydreaming. While daydreaming is often a waste of time, visualization is a very powerful method that can help you become more familiar with a given situation.

For example, when you are giving directions to someone's home, sketching a map of the directions is a great help. A sketch helps *solidify* the directions. In a magazine or a daily newspaper, graphs or other visual tools are used over and over to compare and/or contrast situations and describe them. When you buy something and have to assemble it yourself, the manufacturer's manual usually has pictures as well as word directions. In most sports, especially football and basketball, the coach usually uses a diagram, or drawing, with X's and O's to explain the strategies of a particular play to his players. These are all examples of everyday usage of the strategy of making a drawing when it is not explicitly called for. After all, it has often been said that "one picture is worth a thousand words."

Let us look at a mathematical problem in which you might not initially expect to utilize a visual representation.

> Mr. Adams has two tests in his file of algebra final exams, which he wants to use for two different algebra classes. Each has 26 different questions. He takes the first four questions from test #1 and adds them to the end of test #2. Then he takes the first four questions from test # 2 and adds them to the end of test #1. Each test now has 30 questions. How many questions are the same on both tests?

We can make a drawing, or visual representation, of the situations, both before and after:

Before	Test 1:	A	B	C	D	E	...	W	X	Y	Z				
	Test 2:	1	2	3	4	5	...	23	24	25	26				
After	Test 1	A	B	C	D	E	...	W	X	Y	Z	**1**	**2**	**3**	**4**
	Test 2	1	2	3	4	5	...	23	24	25	26	**A**	**B**	**C**	**D**

The tests now each contain eight questions in common, namely, 1, 2, 3, 4, and A, B, C, D. Although the problem did not specifically call for a visual representation, and clearly the problem could have been solved with other methods, the drawing enabled us to *see* what was going on. It made solving the problem relatively easy. Bear in mind, that when we speak of making a visual representation, it need not be an actual "drawing".

Here is another problem where a visual representation helps us to see what is going on.

> Each side of an equilateral triangle is 40 cm long. The midpoints of each side are joined to form a second equilateral triangle. The midpoints of this triangle are then joined to form a third triangle. We continue joining the midpoints of successive triangles until five triangles have been formed. What is the perimeter of the fifth triangle?

It should be apparent that when a geometric problem is posed — even though it is easily presented verbally — making a diagram of the situation

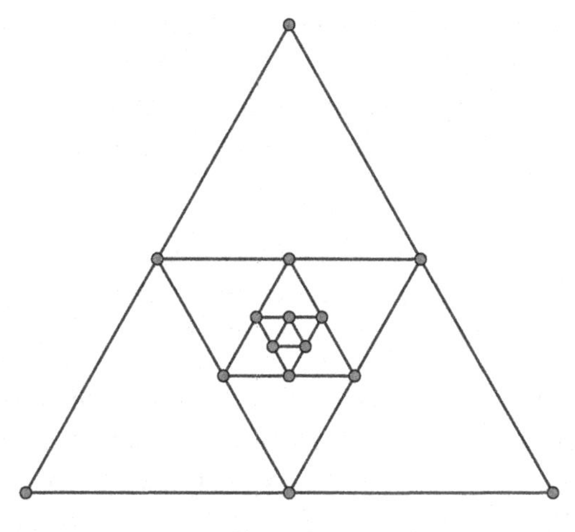

Figure 8.1

being described is useful — if not simply essential! We need to see what is actually being described (see Figure 8.1).

The drawing should remind us of the concept that a line segment joining the midpoints of two sides of a triangle is equal to one-half the length of the third side of the triangle and parallel to it. Thus, each side of any of our triangles is $\frac{1}{2}$ the length of the corresponding side of the previous triangle. The perimeter of each successive triangle is one-half of the perimeter of the previous triangle. For the sake of clarity we will make a table to show the process.

	Side length (cm)	Perimeter (cm)
Triangle #1	40	120
Triangle #2	20	60
Triangle #3	10	30
Triangle #4	5	15
Triangle #5	2.5	7.5

The perimeter of the fifth triangle is 7.5 cm. The drawing we made helped us to visualize the situation and solve the problem. Even though the problem could have been solved without the drawing, seeing it made the solution more easily reached.

To further exhibit the value of the strategy of making a drawing when one is not directly called for in the posed problem, we will consider the following problem:

> At 5:00 o'clock, a clock strikes five chimes in 5 seconds. How long will it take the same clock at the same rate to strike 10 chimes at 10:00 o'clock? (Assume that the chime itself takes no time.)

The answer is *not* 10 seconds! The nature of this problem does not lead us to anticipate that a drawing should be made. However, let us use a drawing of the situation to see exactly what is taking place. In the drawing, each dot represents a chime. Thus, in Figure 8.2, the total time is 5 seconds and there are four intervals between chimes.

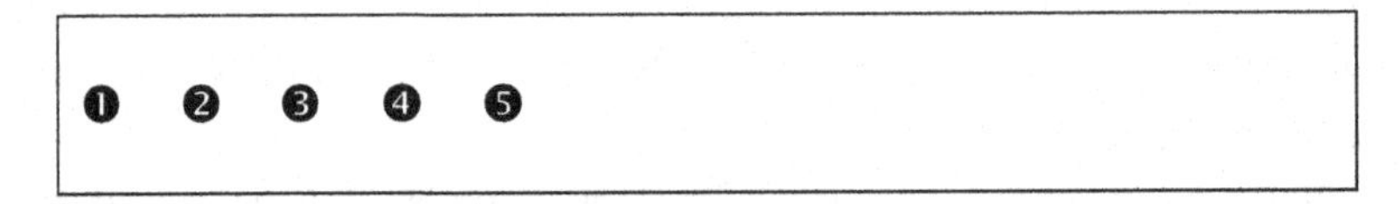

Figure 8.2

Therefore, each interval must take $\frac{5}{4}$ seconds. Now let us examine the second case which we show in Figure 8.3.

Figure 8.3

Here, we can see from the diagram that the 10 chimes require nine intervals. Since each interval takes $\frac{5}{4}$ seconds, the entire clock striking at 10:00 o'clock will take $9 \cdot \frac{5}{4}$ or $11\frac{1}{4}$ seconds.

Using the diagram has made this problem quite simple, which otherwise might have caused some confusion.

This strategy of making a drawing or diagram when one is not called for in the problem is not only a helpful feature in problem solving, but in some cases leads directly to the solution — especially simple problems where the solution might be obvious when a visual representation is made.

PROBLEM 8.1

In Mr. Strauss' classroom, there are 25 seats arranged in five rows with five seats in each row to form a square array. Mr. Strauss decides to change everybody's seat according to the following "rules:" Every student must move either to a seat immediately to his or her left or right, or to a seat immediately in front of them or in back of them. How can this be done?

A Common Approach

The most usual method for attacking this problem is to take 25 markers to stand for the seats and then move them about, according to Mr. Strauss' rules. This is awkward, difficult to keep track of the moves, and probably will not yield the correct answer.

An Exemplary Solution

Instead of trying to move pieces around, let us make a drawing or visual representation of the seating arrangement. We shall draw a representation of the classroom with its 25 seats in a checkerboard fashion as show in Figure 8.4.

Figure 8.4

If the students were to change their seats according to Mr. Strauss' rules, each student would have to move from a shaded seat to an unshaded one, or vice versa. But, there are 13 shaded seats and only 12 unshaded seats. Therefore, it is impossible for students to follow Mr. Strauss' rules.

PROBLEM 8.2

It costs $1.00 to cut and weld a chain link. A woman has seven links and wishes to make a chain. What is the minimum cost to do this?

A Common Approach

The most obvious approach is to open 6 of the links, attach them, and weld them shut. This gives a cost of $6.00. There must be another way to cut the cost.

An Exemplary Solution

We can use our strategy of make a drawing (visual representation):

Open link 2 and connect links 1, 2, and 3 as shown in Figure 8.5.

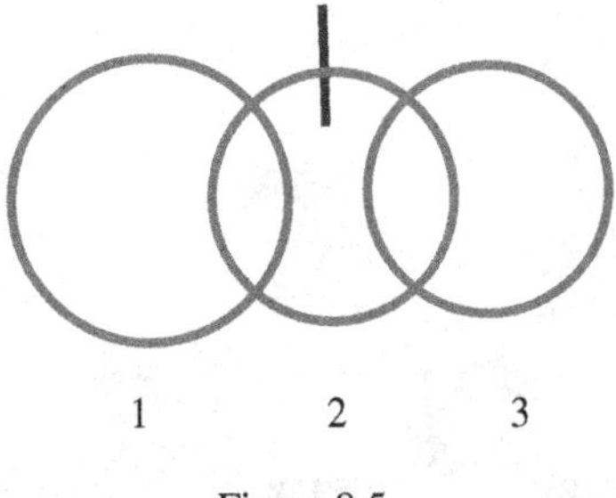

Figure 8.5

Open link 5 and connect links 4, 5, and 6 as shown in Figure 8.6.

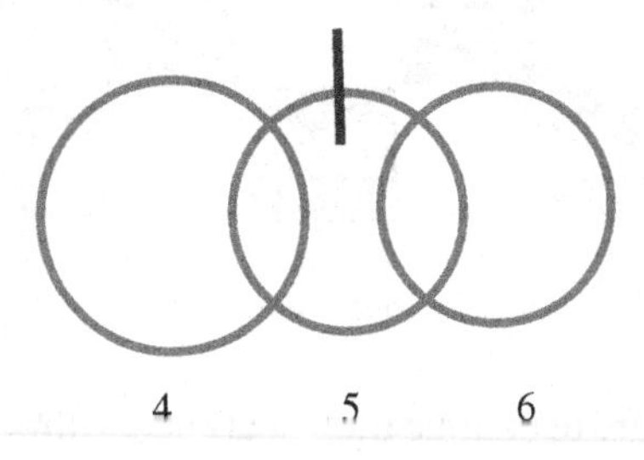

Figure 8.6

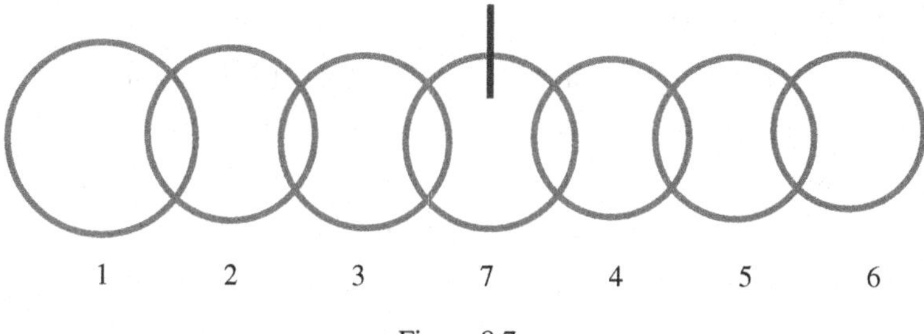

Figure 8.7

Finally, open link #7 and connect the chains: 1–2–3 and 4–5–6 by cutting link 7 as shown in Figure 8.7.

Since we only had to cut three links to make this chain, the cost will be $3 to make the chain.

PROBLEM 8.3

If, on the average, a hen and a half can lay an egg and a half in a day and a half, how many eggs should six hens lay in eight days?

A Common Approach

This is an old problem that has survived the test of time. Traditionally, the problem is solved as follows. Since $\frac{3}{2}$ hens work for $\frac{3}{2}$ days, we may speak of the job of laying an egg and a half ($\frac{3}{2}$eggs) as taking $(\frac{3}{2})(\frac{3}{2})$ or $(\frac{9}{4})$ "hen-days". Similarly, the second job takes $(6)\cdot(8)$ or 48 "hen-days". Thus we form the following proportion:

Let x be the number of eggs laid by 6 hens in 8 days. Then,

$$\frac{\frac{9}{4}\text{ hen-days}}{48\text{ hen-days}} = \frac{\frac{3}{2}\text{ eggs}}{x\text{ eggs}}.$$

Multiplying the product of the means and extremes, we get

$$\left(\frac{9}{4}\right)(x) = 48\left(\frac{3}{2}\right)$$
$$\frac{9x}{4} = 72$$
$$x = 32.$$

An Exemplary Solution

However, as an alternate solution we may set up the following visual representation (here in the form of a tabular layout) of the situation:

$$\frac{3}{2} \text{ hens lay } \frac{3}{2} \text{ eggs in } \frac{3}{2} \text{ days,}$$

Double the previous line:	3 hens lay 3 eggs in $\frac{3}{2}$ days,
Double the previous line:	3 hens lay 6 eggs in 3 days ,
One-third of the previous line:	3 hens lay 2 eggs in 1 day ,
Double the previous line:	6 hens lay 4 eggs in 1 day,
Eight times the previous line:	6 hens lay 32 eggs in 8 days .

Therefore, 6 hens should lay 32 eggs in 8 days.

PROBLEM 8.4

Jack and Sam are both part-time workers in the local pizza shop. The shop is open 7 days a week. Jack works one day and then has 2 days off before he works again. Sam works one day, and then has 3 days off before he works again. Jack and Sam both worked on Tuesday, March 1st. On which other days in March do Jack and Sam work together on the same day?

A Common Approach

A common approach is to begin by making a pair of lists, one for each boy, showing all the dates in March on which each work. Then compare the dates on these lists to determine those on which both boys work. This is a perfectly valid method of solution, and will eventually yield the correct answer.

An Exemplary Solution

A more efficient way to approach this problem is to examine this problem with a visual representation. We will make a drawing of a calendar and then simply place initials on those dates on which each boy works.

S	M	T	W	TH	F	S
		J 1 S	2	3	J 4	5 S
6	J 7	8	9 S	J 10	11	12
J 13 S	14	15	J 16	17 S	18	J 19
20	21 S	J 22	23	24	J 25 S	26
27	J 28	29 S	30	J 31		

Those dates which contain two sets of initials are the dates on which the two boys work together. The figure readily shows these dates to be March 13th and March 25th.

Another clever approach is to attack this problem from another point of view. We know that the numbers 4 and 3 are relatively prime and represent the number of days in the work cycle of each of the boys, respectively. Their common multiple, 12, will provide the days between those on which they work together. Thus, day $1 + 12 = 13$, is a day on which they work together after the first day, and day $13 + 12 = 25$, is the day on which they next work together.

PROBLEM 8.5

At the county fair, there are several employees assigned the task of tracking the number of people who participate in specific activities each day. Rosalinde's notes showed that from Monday through Saturday there were 510 people at the archery range. Gabriel recorded that from Monday through Wednesday, there were 392 people at the archery range. Frank found that on Tuesday and Friday there were 220 people at the archery range. Adele found that on Wednesday, Thursday and Saturday, there were a total of 208 people at the archery range. Finally, Alfred found that from Thursday through Saturday, there were 118 people at the archery range. Assuming that all the figures were correct, how many people were at the archery range on Monday?

A Common Approach

The usual approach is to set up a series of equations using variables to represent the different days of the week. This will result in a set of five linear equations with six variables as follows. Of course, not every variable will occur in every equation.

$$M + T + W + H + F + S = 510, \tag{8.1}$$

$$M + T + W = 392, \tag{8.2}$$

$$T + F = 220, \tag{8.3}$$

$$W + H + S = 208, \tag{8.4}$$

$$H + F + S = 118. \tag{8.5}$$

By solving the set of equations simultaneously, one can attempt to obtain the answer. Once again, the process is rather complicated, and beyond most people's ability. (Few realize that by subtracting equations (8.3) and (8.4) from (8.1), they obtain $M = 82$.)

An Exemplary Solution

Let us make a visual representation (drawing) of the attendance figures as they were reported:

	Monday	Tuesday	Wednesday	Thursday	Friday	Saturday	Total
Rosalinde	×	×	×	×	×	×	510
Gabriel	×	×	×				392
Frank		×			×		220
Adele			×	×		×	208
Alfred				×	×	×	118

Notice that, except for Monday, every day is mentioned three times. This results in twice the attendance being accounted for by the last four people, except for the "missing" Monday. This yields the single equation

$$2(510) - (392 + 220 + 208 + 118) = \text{Monday's attendance}$$
$$1020 - 938 = 82.$$

There were 82 people at the booth on Monday.

PROBLEM 8.6

Amanda, Ian, Sarah, and Emily all entered their pet frogs in the frog jumping contest at the fair, to see whose frog would jump the farthest. Amanda's frog finished ahead of Emily's frog, but not first. Sarah's frog finished behind Amanda's, but was not last. In what order did the jumping frogs finish?

A Common Approach

The most common approach is to take four chips, tokens, or coins to represent the frogs and put a sticker on each with their owners' names. Then move the "frogs" around until the results satisfy the given conditions.

An Exemplary Solution

A simpler approach is to use a visual representation. The first thing we know is that Amanda's frog finished ahead of Emily's frog, but was not first. Our drawing starts like this:

Amanda ← Emily

Sarah's frog was behind Amanda's, but was not last. Continuing the drawing, we get the order of finish as follows:

Ian ←	Amanda ←	Sarah ←	Emily
1	2	3	4

The drawing made it easy to see the order in which the frogs finished.

PROBLEM 8.7

Among 40 boys at Camp Walden, 14 boys swam in the lake, 13 played basketball, and 16 went on a forest hike. Three of the boys played basketball and swam in the lake. Five of them swam in the lake and went on the hike. Eight of the boys played basketball and also were on the hike. Two of them experienced all three events. How many of the boys at this camp were not involved in any of these activities?

A Common Approach

Traditionally, one might begin to solve this problem by adding all the given activities, and then subtracting those duplicate events. Rarely is this procedure effective.

An Exemplary Solution

Let us examine the problem using a visual representation. We will make a drawing to show the data using a Venn diagram (Figure 8.8).

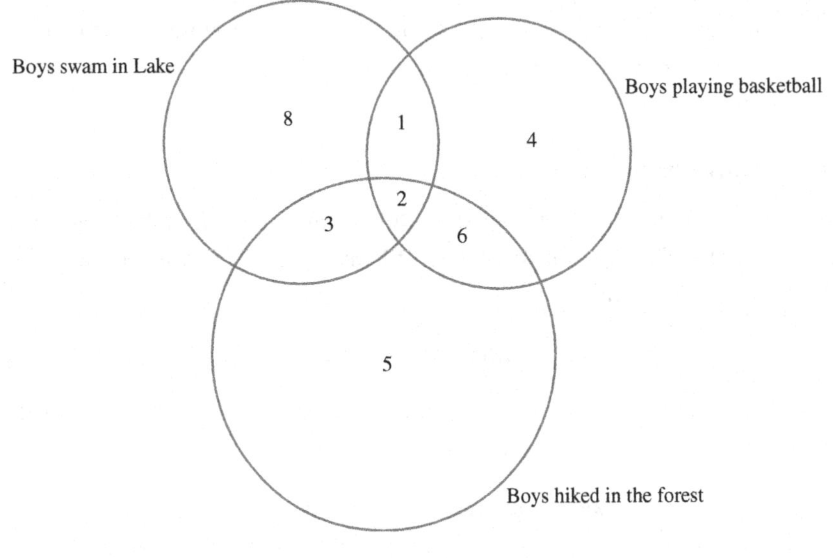

Figure 8.8

The area of overlap for all three circles contains the two boys experiencing all three events. The circles reveal:

Swam in the lake = 14,
Played basketball and took the forest hike = 8,
Swam in the lake and played basketball = 3,
Played basketball = 13,
Swam in the lake and took forest hike = 5,
Took forest hike = 16.

When we add all these individual portions of the Venn diagram, we get $8 + 3 + 2 + 1 + 4 + 6 + 5 = 29$. There were 40 boys in Camp Walden, of whom 29 were involved in the three activities, leaving 11 who participated in none of the three activities.

PROBLEM 8.8

How many integers are there between 4000 and 5000 whose digits are in ascending order?

A Common Approach

One can approach this problem by realizing that the first digit would have to be a 4, which means that the second digit can be any one of the following digits 5, 6, or 7, but cannot be an 8 or 9, since then they would not be enough options for the rest of the number to allow for an ascending order of digits. With some logical thinking in this fashion the following should result. The required numbers are: 4567, 4568, 4569, 4578, 4579, 4589, 4678, 4679, 4689, and 4789.

An Exemplary Solution

To approach this problem in perhaps a more organized fashion using a diagram, even though it is not called for from the nature of the problem, we will use a tree diagram as shown in Figure 8.9.

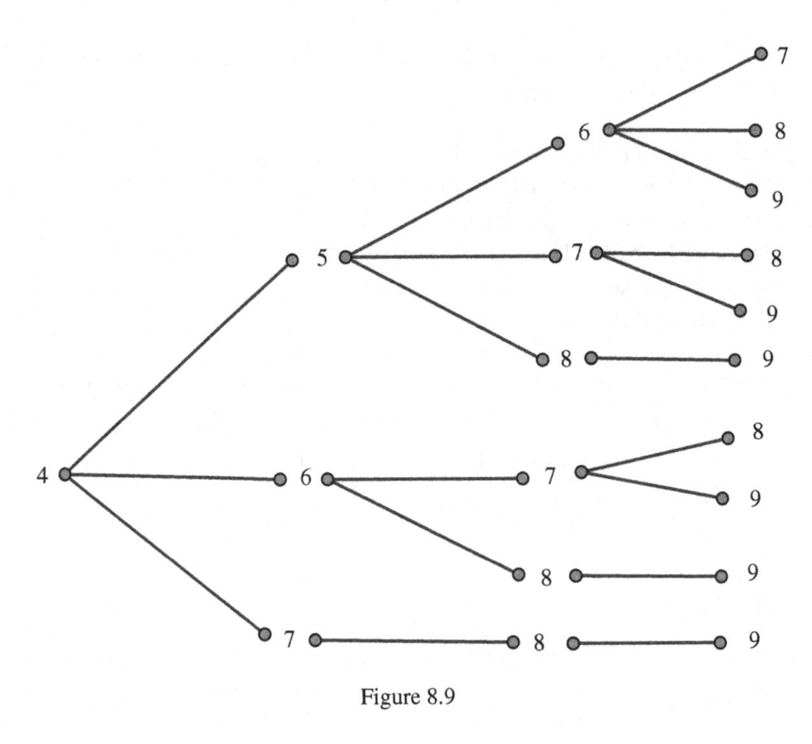

Figure 8.9

Beginning with the numeral 4, each path leads to a number which fits in the range from 4000 to 5000. There are 10 such paths which generate the following numbers: 4567, 4568, 4569, 4578, 4579, 4589, 4678, 4679, 4689, and 4789. In this way we have organized the numbers required using a diagram when one was not called for.

PROBLEM 8.9

My brother has a figurine collection of two-legged apes and four-legged buffalos. If there are 100 figurines in all, and a total of 260 legs, how many of each figurine does he have?

A Common Approach

The most often-used approach is to solve a pair of equations simultaneously. Let a stand for the number of ape figurines and b stand for the number of buffalo figurines. Then we obtain the following equations:

$$a + b = 100,$$
$$2a + 4b = 260.$$

Multiplying the first equation by 2 we obtain

$$2a + 2b = 200,$$
$$2a + 4b = 260.$$

Subtracting the above two equations, we get

$$2b = 60,$$
$$b = 30.$$

There are 30 buffalo figurines and 70 ape figurines.

An Exemplary Solution

Let us make use of our visual representation (make a drawing) to resolve the problem. First, let us reduce the numbers in the problem by a factor of 10 to make the problem more manageable (but we must remember to multiply our result by 10 to get back to the original numbers) so that we then have 26 legs and 10 animals. Then we shall draw ten circles to represent the 10 animals. Now, regardless of whether the animal is an ape or a buffalo, it has to have at least two legs (Figure 8.10).

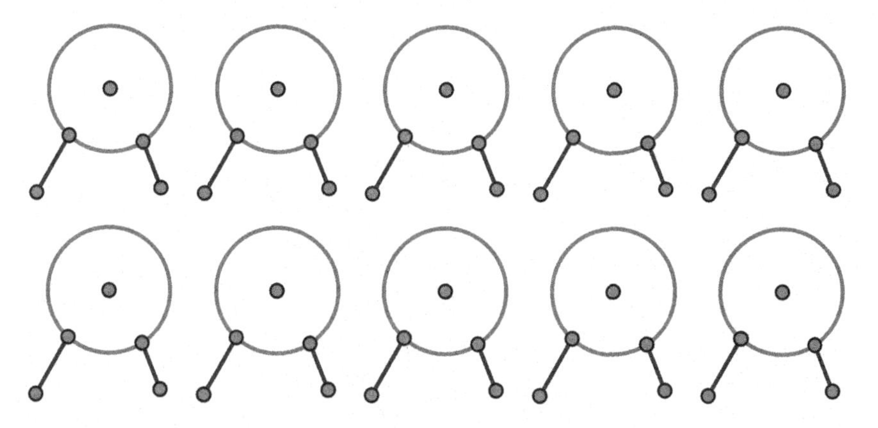

Figure 8.10

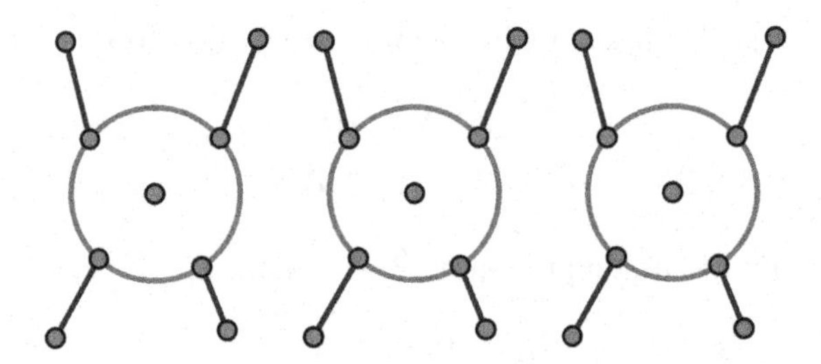

Figure 8.11

There now are six legs not accounted for — they must be installed by 2's (Figure 8.11).

There are three four-legged animals leaving seven two-legged animals. We must multiply by 10, so we get 30 buffalo figurines and 70 ape figurines.

Chapter 9

Accounting for All Possibilities

We know that organizing the data in a problem can sometimes be very revealing. When we look for a pattern, for example, carefully organized data in a list or table can aid in the search and help us discover the pattern. However, one particular type of list that is extremely important is the *exhaustive* list. In this kind of organized list, all the possibilities are listed in some systematic way. Somewhere included on this list will be what we are looking for. Constructing an exhaustive list enables us to examine all the possibilities in a carefully-organized manner.

As an example, suppose you have a lamp that is just not working. We can list all the possibilities. (Of course, it might be a mental list, but it *is* a list, nevertheless.) The problem might be caused by a burned-out light bulb, a bad wire, an electric outlet that is not working, a circuit breaker that may have been tripped, or even just a bad switch on the lamp. One by one we can eliminate each item until we come to the one that is the cause of the malfunction. A mathematical example might be the following:

> We start with a 2-digit perfect square. When we insert a single digit between the two, we obtain a 3-digit perfect square. What are the 3-digit squares we have made?

Let us examine all the possibilities. First of all, we will make an exhaustive list of all the 2-digit perfect squares; there are six.

$$16, 25, 36, 49, 64, 81.$$

Now we can make an exhaustive list of all the 3-digit perfect squares:

$$100, 121, 144, 169, 196, 225, 256, 289, 324, 361, 400, 441, 484, 529, 576, 625, 676, 729, 784, 841, 900, 961.$$

Using the second list, we examine the first and third digits of each of these to see which could have been made from the 2-digit squares. We find that **196** (inserting 9 between 1 and 6 in 16), **225** (inserting 2 between 2 and 5 in 25), and **841** (inserting 4 between 8 and 1 in 81), are the only ones that satisfy the given conditions. The two exhaustive lists showed us all the possibilities. Notice that an exhaustive list will not only contain the answer to the problem, but it also limits the number of possibilities to be investigated.

Here is another example of using this valuable strategy.

> Two people are sitting on a park bench. One of the two people is female. What is the probability that they are both females?

We will make a list, accounting for all the possibilities (M = male, F = female):

$$\text{M–M} \quad \text{M–F} \quad \text{F–M} \quad \text{F–F}.$$

The list shows four possibilities, but in this problem, the first one, M–M, need not be considered, since we know at least one person is female. So there are really only three possibilities to examine. Thus, only one of the three choices can be two females, or to answer the question, the probability that both are females is $\frac{1}{3}$.

To get a better understanding of the value of this problem-solving technique, we will consider another example:

> The local movie theater is showing a set of cartoons in each of two theaters in the morning. The shows must both end by 1:00 p.m., so as to begin their feature films. In theater A, the cartoon show starts at 9:00 a.m., starts over at 9:28, and repeats every 28 minutes thereafter. In theater B, the cartoon show starts at 9:00 a.m., but starts over every 35 minutes thereafter. Joanne wants to see both shows. At what times do the two shows next start together?

We can make an exhaustive list for the times the shows start in both theaters.

Theater A	Theater B
9:00	9:00
9:28	9:35
9:56	10:10
10:24	10:45
10:52	**11:20**
11:20	11:55
11:48	12:30
12:16	
12:44	

Any further starting times would be after 1:00 p.m. We have listed them all! Somewhere on these lists of all the possibilities is the answer. The lists show us that the only other time the two shows start together is 11:20 a.m.

The strategy is a very valuable one, but you must be certain that you have included *all* the possibilities! We need a carefully organized approach to make certain we have them all. As with all the strategies one has to select from, one must be careful to use the appropriate one. In the case of the strategy of considering all possibilities, the solution can become more visible.

PROBLEM 9.1

A mathematics teacher noted that his present age is a prime number. He observed the next time his age would be a prime number was as far in the future as the previous prime number age was in the past. How old is the mathematics teacher?

A Common Approach

This problem does not lend itself to very many alternative methods of solution. Typically, one begins to test various numbers with the hope of "stumbling upon the right number".

An Exemplary Solution

Here we will surely appreciate the strategy of accounting for all possibilities. We consider the following list:

Primes:	2	**3**	**5**	**7**	11	13	17	19	23	29	31	37	41
Differences:	1	**2**	**2**	4	2	4	2	4	6	2	6	4	2
Primes:	43	**47**	**53**	**59**	61	67	71	73	79	83	89	97	101
Differences:	4	**6**	**6**	2	6	4	2	6	4	6	8	4	

In doing so, a list of primes from 1 to 100 (although the mathematics teacher's age would probably only require inspecting the prime numbers between 20 and 80) reveals only two situations where three consecutive primes have a common difference. The first case, 3, 5, and 7, could not possibly be used since the mathematics teacher's age cannot be 5 years. The second case, 47, 53, and 59, seems to suggest a reasonable age range. Thus, the mathematics teacher's age would be 53.

PROBLEM 9.2

Find the number of ways in which 20 coins, consisting of nickels, dimes and quarters, can total \$3.10.

A Common Approach

It is expected that one might immediately try to create algebraic expressions, which reflect the given information in the problem statement. Therefore, they will get: $n + d + q = 20$, where n, d, and q represent the number of nickels, dimes, and quarters, respectively. This can be written as $n = 20 - q - d$. Furthermore, $25q + 10d + 5n = 310$, which by combining the last two equations gives us: $25q + 10d + 5(20 - q - d) = 310$. Then $4q + d = 42$, or $q = 10 + \frac{2-d}{4}$. At this point the common approach is to try various values to determine which work best.

An Exemplary Solution

We are now at a point at which we can use a clever method, that of accounting for all possibilities for the value of d. First, we notice that we must be certain that q is an integer. That would require that we isolate the fractional part of q, that is $\frac{2-d}{4} = k$, or $d = 2 - 4k$. By substituting above, we get $q = 10 + k$, and $n = 20 - q - d = 20 - (10 + k) - (2 - 4k)$, or $n = 8 + 3k$.

Since $d = 2 - 4k$, the value of k would have to be either zero or negative.

The chart below shows the various value possibilities for k and the ensuing values for d, q, and n.

k	d	q	n
0	2	10	8
−1	6	9	5
−2	10	8	2
−3	−10	7	−1

All seems to be fine with $k = 0, -1, -2$. However, when $k = -3$, then $d = 2 - 4(-3) = 14$, and $n = 8 + 3(-3) = -1$, which are meaningless in this problem. Thus, we have the answer to our problem, namely, there are three ways to make a total of \$3.10.

PROBLEM 9.3

In order to ship boxes of tuna fish cans, the company can pack a small carton that holds eight cans, or a large carton that holds ten cans. They always try to use as many large cartons as possible, for cost-efficient reasons. If an order for 96 cans of tuna fish is received, how should they pack them for shipping?

A Common Approach

The problem lends itself to an interesting mathematical solution. If we let x represent the number of small cartons, and y the number of large cartons, we get the equation:

$$8x + 10y = 96.$$

However this is one equation with two variables, a situation that usually leads to multiple answers. Since the values of x and y must be integral, the equation is referred to as a Diophantine Equation, after the Greek mathematician, Diophantus (ca. 208–292 AD). Let us see if we can solve it. Solving for x in terms of y yields the following:

$$x = \frac{96 - 10y}{8}$$
$$= 12 - \frac{10y}{8}$$
$$= 12 - y - \frac{2y}{8}.$$

But $-\frac{2y}{8}$ must be integral in order for x to be a whole number. Let $y = 4$. Then $\frac{-2y}{8} = -1$, and $x = 12 - 4 - 1 = 7$. Thus, we have seven small cartons and four large ones. Are there other answers? Let us see if we can find any other answers. In a similar manner, we can obtain 12 and 0 by letting $y = 0$. Finally, letting $y = 8$, we get $x = 2$.

An Exemplary Solution

The strategy that best fits this problem is to attempt to resolve it by accounting for all possibilities and organizing the data with a table.

Small carton	No. of cans	Large cartons	No. of cans	Total cans
12	96	0	0	96

We seem to have gotten one set of answers right away! This satisfies the problem's *numerical* conditions — that they must ship 96 cans. However, is this the only possibility? After all, this answer says they do not ship *any* large cartons. Since they try to use as many large cartons as possible, this answer seems odd. Let us continue the table and try to find all the possibilities.

Small carton	No. of cans	Large cartons	No. of cans	Total cans
12	**96**	**0**	**0**	**96**
11	88	No way to pack the remaining 8 cans		
10	80	No way to pack the remaining 16 cans		

(Continued)

(*Continued*)

Small carton	No. of cans	Large cartons	No. of cans	Total cans
9	72	No way to pack the remaining 24 cans		
8	64	No way to pack the remaining 32 cans		
7	**56**	**4**	**40**	**96**
6	48	No way to pack the remaining 48 cans		
5	40	No way to pack the remaining 56 cans		
4	32	No way to pack the remaining 64 cans		
3	24	No way to pack the remaining 72 cans		
2	**16**	**8**	**80**	**96**

There are three possibilities: 2 small cartons and 8 large, 7 small cartons and 4 large, and 12 small cartons and 0 large. However, since they wish to use as many large cartons as possible, the *answer* to the problem is two small cartons and eight large cartons. Notice that from a mathematical viewpoint, all three answers satisfy the given condition of 96 cans to be shipped. However, the context of the problem eliminates two of the three answers that the table so nicely revealed.

PROBLEM 9.4

On a standard die, the dots on the opposite faces have a sum of 7. How many different sums of dots on three adjacent faces are there on this standard die?

A Common Approach

Typically one would try to draw a die, and then systematically count the dots on adjacent faces to come up with an answer. Others will try to list all possible combinations of dots on any three faces with no regard for whether or not they are adjacent.

An Exemplary Solution

We will organize the data in such a way that we can account for all possibilities. Since the sum of opposite faces is 7, the only possibilities for this to happen would be:

1 and 6,

2 and 5,

3 and 4.

Now, we know that if we consider three adjacent faces, they must share a common vertex. Since there are eight vertices, there will be eight sets of three adjacent faces. We now check to see whether or not they all have different sums. To do this, we will select all possible sets of three by choosing one number from each of the three pairs of opposite faces described above, and then take their sum. To be certain that we have all the possibilities, we will select them in an organized manner:

$\{1, 2, 3\}$; sum $= 6$ $\{1, 5, 3\}$; sum $= 9$ $\{6, 2, 3\}$; sum $= 11$
$\{6, 5, 3\}$; sum $= 14$ $\{1, 2, 4\}$; sum $= 7$ $\{1, 5, 4\}$; sum $= 10$
$\{6, 2, 4\}$; sum $= 12$ $\{6, 5, 4\}$; sum $= 15$

There are eight *different* sums, as we expected with the eight vertices.

PROBLEM 9.5

During the last census, a man told the census taker that he had three children. When he was asked for their ages, he replied, "I cannot tell you that, but I will tell you the product of their ages is 72. Furthermore, the sum of their ages is the same as my house number." The census taker ran to the front of the house and looked at the house number. "I still cannot tell," she said. The man answered, "Oh yes, I forgot to tell you that my oldest child loves blueberry pancakes." The census taker promptly wrote down their ages. How old are they? (Consider only whole number ages.)

A Common Approach

The most common approach is to try to form a series of equations. If we let the ages of the three children be x, y, and z, we obtain

$$x \cdot y \cdot z = 72,$$

$$x + y + z = h \quad \text{(where } h \text{ is the house number).}$$

This leaves us with a rather unmanageable situation: we now have a system of two equations in four variables. It seems to be impossible to solve the problem. We could guess, but that might take a long time to arrive at the answer.

An Exemplary Solution

Let us use our strategy of accounting for all possibilities. Since the product of their ages is 72, we will start by listing all the number triples whose product is 72. We will use our strategy of organized data to make sure we have all the possibilities.

1, 1, 72	1, 4, 18	2, 2, 18	2, 6, 6
1, 2, 36	1, 6, 12	2, 3, 12	3, 3, 8
1, 3, 24	1, 8, 9	2, 4, 9	3, 4, 6

Notice that after 1, 8, 9 the triples "double back". This is the complete set of triples whose products are 72. Somewhere on this list lurks the answer. Now we also know that the sum of their ages is my house number:

$1+1+72=74$	$1+4+18=23$	$2+2+18=22$	$\mathbf{2+6+6=14}$
$1+2+36=39$	$1+6+12=19$	$2+3+12=17$	$\mathbf{3+3+8=14}$
$1+3+24=28$	$1+8+9=18$	$2+4+9=15$	$3=4+6=13$

The census taker saw the man's house number, yet she couldn't tell the ages. Why not? If, for example, the house number had been say, 18, then the ages would be easily known as 1, 8, and 9. However, her inability to determine which of these triples is the right one, must have been because there are two sets of triples whose sum is 14, the house number must have been 14. However, when the man said "My *oldest* loves blueberry pancakes," the census taker knew that there had to be an *oldest*. The ages must have been 3, 3, and 8. Since the triple 2, 6, and 6 does not have an *oldest*.

Notice that the blueberry pancakes are really just a distractor. It is the word "oldest" that provides the key to the problem.

PROBLEM 9.6

In Figure 9.1, what is the number of common tangents to exactly two circles at a time?

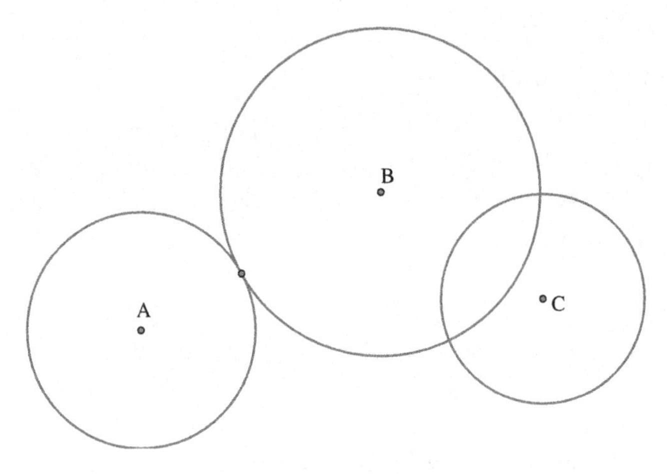

Figure 9.1

A Common Approach

While we could draw all of the common tangents and count them, we would not necessarily get them all, since the drawing would probably become too confusing.

An Exemplary Solution

An organized way to solve this problem is to take the circles two at a time, and account for all the possibilities.

Circles A and B: 2 external tangents + 1 internal tangent
Circles A and C: 2 external tangents + 2 internal tangents
Circles B and C: 2 external tangents.

Thus there are a total of nine tangents in all. The problem was easily resolved by accounting for all the possibilities.

PROBLEM 9.7

Maria is helping her father tile a rectangular playroom floor. They use exactly 2005 square tiles. Some are black and some are white. The border is one tile in width, and consists of only black tiles. The rest of the tiles are white. How many white tiles did they use to make the playroom floor?

A Common Approach

If we draw the figure, we obtain a pair of rectangles as shown in Figure 9.2. If the dimensions of the inner rectangle are x and y, then the width of the outer rectangle is $x + 2$, and the length is $y + 2$.

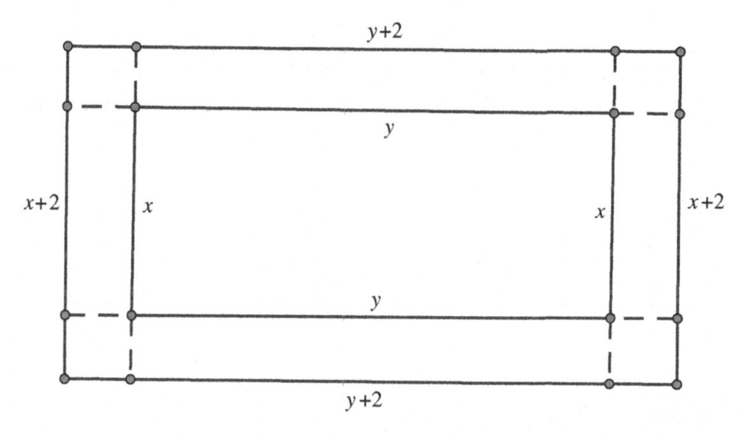

Figure 9.2

It is to be expected that a natural reaction is to represent the given information algebraically by forming the following equation:

$$(x + 2)(y + 2) = 2005.$$

By multiplying and simplifying this equation we get the following:

$$xy + 2y + 2x + 4 = 2005$$
$$xy + 2y + 2x = 2001.$$

We are now faced with one equation and two variables, and what we seek is xy. This leads to a dilemma, and not a practical solution.

An Exemplary Solution

With some logical reasoning, we inspect the given information from another point of view, namely, considering all possibilities. The number of tiles, 2005, can only be factored in two ways: either $1 \cdot 2005$ or $5 \cdot 401$. This presents us with two possible dimensions for our desired rectangle. The first situation can be disregarded, because there would be no white tiles at all, if the width were 1. So, the playroom floor must consist of 5 by 401 tiles. Since the outer "frame" is one-square tile all the way around, the dimensions of the inner rectangle, made up of all white tiles, contain two fewer tiles in each direction, one in each corner. When we remove the two tiles from each dimension, the number of white tiles remaining for the inner rectangle is $(3 \cdot 399)$ or 1197. Therefore, they used 1197 white tiles to cover the playroom floor.

PROBLEM 9.8

Given the integers from -100 to $+100$, how many of these integers squared have a unit digit of 1?

A Common Approach

A natural reaction is to begin by listing all the integers from 1 to 100. Then square each one and count those that have 1 as the unit. Then double this number to account for the -1 to -100 integers.

An Exemplary Solution

Let us use our accounting for all the possibilities strategy. The only numbers whose square has a units digit of 1 are the numbers whose units digit is either a 1 or 9. Thus there are exactly 20 possibilities, namely 1, 11, 21, 31, 41, 51, 61, 71, 81, 91, 9, 19, 29, 39, 49. 59, 69, 79, 89, and 99. Doubling — to include the negative numbers — we have all the possibilities, namely, 40 such integers within the given range.

PROBLEM 9.9

Figure 9.3 below shows three faces of a cube. If the six faces of the cube are numbered consecutively, what is the sum of the numbers on all six faces?

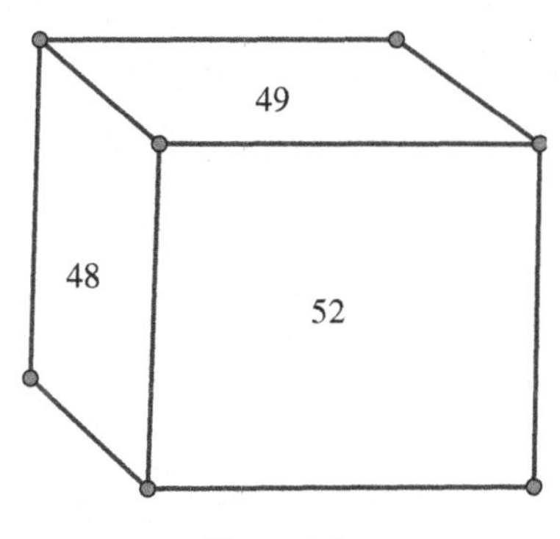

Figure 9.3

A Common Solution

Most people will observe that the numbers shown on the faces of the cube begin with 48 and 49. The most usual assumption is to merely continue the sequence for six terms, arriving at 48, 49, 50, 51, 52, 53 as the numbers on the faces. Since the third face shown, namely 52, appears in their sequence, some folks are usually quite content and give the sum of these six numbers, 303, as their answer.

An Exemplary Solution

However, some problem solvers have not accounted for all the possibilities. We see three of the six faces. Since we see 48, 49, and 52, there must also be a 50 and a 51. However, the sixth number could occur on either end of the sequence. Thus there are *two* possibilities for the sixth number, either 47 or 53. This yields two possible sums, 297 or 303.

Chapter 10

Intelligent Guessing and Testing

Somehow, the very idea of guessing as a problem-solving strategy raises some eyebrows. Indeed, many of us can recall a teacher saying to someone who had volunteered an unusual answer, "Do you *know*, or are you just *guessing*?" In some books, this method of guessing and testing is sometimes referred to as the method of "trial and error," which is sometimes considered a more negative expression. The added adjective, *intelligent* guessing and testing should put your mind at ease, and reassure you that this is, indeed, a viable, and even an often useful strategy.

We have been using the guess-and-test strategy most of our lives. For example, when cooking, we make a guess as to how well-done a roast in the oven might be. Then we use a meat thermometer to test our guess to see if we were right. If not, we put the roast back in the oven, and let it cook a while longer before repeating the process. When we are trying to find a particular place while driving, we "guess" it will be on a specific street. If it turns out it is not there, we make another guess based on what information we received from our first trial.

In problem solving, when the general case in a particular problem is too complicated, we can use this strategy to narrow down the generalities with specific guesses. As we test each of our guesses, we find information enabling us to refine the next guess, and leading us towards an ultimate answer to the problem.

When used to help solve problems, guesses should not be made randomly, nor made wildly with no apparent justification. After reading a problem carefully, we decide on a possible approach, and, if appropriate, we make a guess. We then test this guess based upon the given conditions of the problem. If the problem is not solved, we then make a second guess, based upon the information we obtained from the previous guess. We then test this latest guess. The process can be continued, each time refining our subsequent guesses based on the information from previous guesses and tests until we have enough information to resolve the problem situation. For example, suppose we are asked to find the next two terms in the sequence 2, 0, 4, 3, 6, 7, 8, 12, 10, 18, ______, ______. What do we notice? It appears that there may be terms increasing and decreasing in a random sort of way. Perhaps there are two sequences, interwoven. This seems to be an *intelligent* guess. Let us test it!

Sequence #1:	2	4	6	8	10	(the odd-place terms),
Sequence #2:	0	3	7	12	18	(the even-place terms).

Looks like our guess was correct; there are indeed two interwoven sequences. Sequence #1 consists of the even numbers. The next term in this sequence would be 12. Sequence #2 shows that the differences between successive terms are increasing by 1 each time, to be differences of 3, 4, 5, 6, and so on. The next term would be 25. Thus we have the answer to our problem: the next two terms are 12 and 25. It is important for the reader to notice that we were also making use of our strategy of finding a pattern. It is not uncommon that more than one strategy can be involved when solving a problem.

Notice too, that our guesses were not made in a wild, unknowing manner. Rather they were all based upon careful observation of what was given and what we wanted to find. These guesses were made intelligently! Keep in mind that this strategy is referred to as *intelligent* guessing and testing for a reason. It is a very useful strategy.

We will consider another use of this strategy with the following problem.

> A local firm has to fill an order for rubber marbles and rubber balls. Rubber balls weigh one ounce each, while the rubber marbles, being solid, weigh two ounces each. Both are exactly the same size. A box can hold 50 of these spherical objects. The best rate for shipping is for the box to hold exactly 80 ounces. How many of each should they put in a box?

An interesting approach, rather than using algebra, is to use the intelligent-guessing-and-testing strategy. We can make a table to keep track of our guesses. We will start in the middle, with 25 of each.

Rubber marbles	(ounces)	Rubber balls	(ounces)	Total	(ounces)
25	(50)	25	(25)	75	(too little)
35	(70)	15	(15)	85	(too much)
30	(60)	20	(20)	80	YES!

They should package 30 rubber marbles and 20 rubber balls. If we had tried other values, we would notice the tendency toward the above correct solution. Yet, intelligent guessing has allowed us to limit our guesses.

Let us look at another problem whose solution can profit nicely by using this strategy.

> Tossing darts is a very popular sport in many countries. Pamela threw some darts at a dartboard, whose sections are labeled 2, 3, 5, 11, and 13. If she scored exactly 150, what is the fewest number of darts Pamela could have thrown?

Since we want to find a minimum number of darts, we should try to have as many of the higher-valued sections as possible. Let us make a series of guesses together with a table to organize our data.

13	11	5	3	2	Number of darts	Total score
12					12	156 (too big)
11		1		1	13	150
10		4			14	150
9	**3**				**12**	**150**
8	4			1	13	150
7		11		2	20	150

The fewest number of darts she could have thrown was 12. Again, notice that we used an organizing-data strategy to help keep track of our guesses.

A table is often a very valuable aid when using the intelligent-guessing-and-testing strategy, since it helps keep track of the information we obtain from each successive guess.

PROBLEM 10.1

A local farm has a field of blueberry plants in a square array so that the number of rows equals the number of columns. The farmer decides to increase the size of the blueberry field equally in the number of rows and columns. His new field contains 211 additional blueberry plants. How many plants did he have in one row of the original field?

A Common Approach

We can use algebra to write an equation. Let x represent the number of rows and the number of columns. So the original number of plants is $x \cdot x$ or x^2. If we let b represent the additional plants for each row and column, the new number of plants is $(x + b)^2$. Now we have the equation

$$x^2 + 211 = (x + b)^2$$
$$x^2 + 211 = x^2 + 2bx + b^2$$
$$211 = b^2 + 2bx.$$

But this poses a problem. We have only one equation, a quadratic in b but also containing x. What can we do now? Perhaps we can substitute values for the variables and see if we can solve the equation. Although this may lead to a correct answer, it is not a very efficient method.

An Exemplary Solution

Let us use our intelligent-guessing-and-testing strategy to complete the solution. It happens that the number 211 is a prime number, and x and b must be whole numbers. If we factor the equation from above, we obtain

$$211 = b(b + 2x).$$

Since 211 is a prime, it has only two factors: 211 and 1. Thus b must equal 1, and $b + 2x$ must equal 211. So, $2x = 210$ and $x = 105$. There are 105 plants in one row of the original field.

PROBLEM 10.2

Jack wants to fence in a rectangular garden plot for vegetables. He has 20 feet of fencing to use. What will the dimensions be if he wants to enclose the largest possible area?

A Common Approach

The most instinctive approach is algebraic. We can try to formulate equations and solve them simultaneously. We would then let x represent the length, and y represent the width. Then we obtain:

$$2x + 2y = 20 \quad \text{or} \quad x + y = 10.$$

Formulating a second equation brings in a new problem — how can we represent a *maximum* area? That is, we want xy = a maximum. What can we do? Looks like we might have to abandon this approach.

An Exemplary Solution

An immediate guess shows that a length of 8 and a width of 2 "works". But so do other pairs of numbers. Let us use our strategy of intelligent guessing and testing to see which dimensions give the largest area. We will use a table to keep track of our guesses. Since we multiply one width by one length to get the area, we make use of the semi-perimeter of 10 in our table. Start with the largest possible length.

Length	Width	Arca
9	1	9
8	2	16
7	3	21
6	4	24
5	**5**	**25**
4	6	24
3	7	21
2	8	16
1	9	9

Looks like a 5×5 rectangle (square) has the greatest area. But what about fractional dimensions? The problem did not say the dimensions had to be integral. Let us put some fractions into our table and see what happens.

Length	Width	Area
9	1	9
8	2	16
7	3	21
6.5	3.5	22.75
6	4	24
5.5	4.5	24.75
5	**5**	**25**
4.5	5.5	24.75
4	6	24
3	7	21
2	8	16
1	9	9

It appears that the rectangle of largest area with the perimeter of 20 feet is a $5' \times 5'$ square. Some people already are familiar with the fact that the square has the maximum area for a rectangle with a given perimeter. If that is the case, then the answer is quickly found as the square of perimeter 20 and thus, an area of $5 \times 5 = 25$ square feet.

PROBLEM 10.3

What is the smallest prime number greater than 510? (Remember a *prime number* is a number with exactly two divisors: 1 and the number itself.)

A Common Approach

Since the problem asks for the smallest prime number greater than 510, we begin with 511, then 512, and so on trying each successive number to find divisors up to one-half of the number being tested. When none of these possible divisors work, then we know a prime number has been found.

An Exemplary Solution

Let us use our intelligent guessing and testing strategy to narrow down the possible choices. We know that numbers greater than 510 cannot be prime if their units digit is 0, 2, 4, 5, 6, or 8. Furthermore, you may recall that a number with a digit-sum of 3 is divisible by 3. This will eliminate some potential numbers greater than 510, such as 513. So, we restrict our guesses to 511, 517, 521, and so on. We find that 521 is the next larger prime.

PROBLEM 10.4

The one-mile relay team consists of four runners: Gustav, Johann, Richard, and Wolfgang. Coincidentally, the order in which they run their quarter-mile lap is the same as the alphabetical order of their names. Each runner runs his quarter-mile lap 2 seconds faster than the previous runner. They finished the race in exactly 3 minutes 40 seconds. How fast did each runner run his lap?

A Common Approach

With some simple algebra we can solve the problem as follows:

$$\begin{aligned} x &= \text{the time it takes Gustav to run his lap;} \\ x - 2 &= \text{the time it takes Johann to run his lap;} \\ x - 4 &= \text{the time it takes Richard to run his lap;} \\ x - 6 &= \text{the time it takes Wolfgang to run his lap.} \end{aligned}$$

$$x + (x - 2) + (x - 4) + (x - 6) = 220$$

$$(\text{3 minutes 40 seconds} = \text{220 seconds})$$

$$4x - 12 = 220$$

$$4x = 232$$

$$x = 58.$$

It takes Gustav 58 seconds to run his lap, Johann takes 56 seconds, Richard takes 54 seconds, and Wolfgang takes 52 seconds.

An Exemplary Solution

Of course, this solution depends upon having a knowledge of algebraic equations. However, we can still solve the problem by making use of the intelligent-guessing-and-testing strategy. We shall assume that the runners run at approximately the same speed; thus, we can divide 220 by 4, and obtain 55 as our first guess.

	Gustav	Johann	Richard	Wolfgang	Total
Guess #1	55	53	51	49	208 (Too small)
Guess #2	60	58	56	54	228 (Too much)
Guess #3	59	57	55	53	224 (Too much)
Guess #4	58	56	54	52	220 (Correct)

Thus, Gustav took 58 seconds, Johann 56 seconds, Richard 54 seconds, and Wolfgang 52 seconds.

PROBLEM 10.5

Dan has a box that contains only 13¢ stamps and 8¢ stamps. He wishes to mail a package that costs exactly $1 in postage. How many of each denomination stamp will Dan put on the package, if he uses only his 13¢ and 8¢ stamps?

A Common Approach

We can attempt to use our algebraic skills to solve this problem. If we let x represent the number of 13¢ stamps and y represent the number of 8¢ stamps, we obtain the following equation:

$$0.13x + 0.08y = 1.00.$$

If we change everything to cents have:

$$13x + 8y = 100.$$

But this is one equation in two variables, which suggests multiple answers. However, since the number of stamps must be integral, we are attempting to solve, what is called a Diophantine equation.

We begin by solving for y: $y = \frac{100-13x}{8}$. When we do the indicated division and separate the quotients and remainders, and then write the remainders together, we get: $y = 12 - x + \frac{4-5x}{8}$.

But, the fractional part must be a whole number, since we cannot have a fractional number of stamps. Let us choose a value for x that makes the fractional part an integer. Let $x = 4$. Then $y = 12 - 4 + (-2)$, or $y = 6$.

Dan will, therefore, use 6 of the 8¢ stamps and 4 of the 13¢ stamps. (But are there other possibilities? Have we found *all* the possible answers?)

An Exemplary Solution

A more elegant approach is to use our strategy of intelligent guessing and testing by using a table to organize the data.

Number @ 13¢	Value	Number @8¢ value	Total value
7	91¢	Cannot make remaining 9¢ with 8 stamps	
6	78¢	Cannot make remaining 22¢ with 8 stamps	
5	65¢	Cannot make remaining 35¢ with 8¢ stamps	
4	**52¢**	**and 6 of the 8¢ is 48¢,**	**giving a total of $1**
3	39¢	Cannot make the remaining 61¢ with 8¢ stamps	
2	26¢	Cannot make the remaining 74¢ with 8¢ stamps	
1	13¢	Cannot make the remaining 87¢ with 8¢ stamps	
0	13¢	Cannot make the remaining $1.00 with 8¢ stamps	

Thus 4 of the 13¢ stamps and 6 of the 8¢ stamps will make exactly the $1.00 that Dan needs. Notice that the table reveals clearly that this is the only possible answer.

PROBLEM 10.6

Two positive integers differ by 5. If their square roots are added, the sum is also 5. What are the two integers?

A Common Approach

The traditional approach is to set up a system of equations as follows:

Let $x =$ the first integer.
Let $y =$ the second integer.

Then,

$$y = x + 5$$
$$\sqrt{x} + \sqrt{y} = 5$$
$$\sqrt{x} + \sqrt{x+5} = 5.$$

Squaring both sides,

$$x + x + 5 + 2\sqrt{x(x+5)} = 25.$$

Simplifying,

$$2\sqrt{x(x+5)} = -2x + 20.$$

Squaring again,

$$4x^2 + 20x = 4x^2 - 80x + 400$$
$$100x = 400$$
$$x = 4$$
$$y = 9.$$

The two integers are 4 and 9.

An Exemplary Solution

Obviously, this procedure requires a knowledge of equations with radicals, and requires a great deal of careful algebraic manipulation. As an alternative let us make use of our intelligent-guessing-and-testing strategy to solve this problem. Since the sum of the square roots of the two integers is 5, the individual square roots must be 4 and 1, or 3 and 2. Thus the integers must be 16 and 1, or 9 and 4. However, only when we consider the determined difference of the squares, do we know that since 9 and 4 have a difference of 5 that must be the correct answer.

PROBLEM 10.7

The coach of the soccer team lets her players choose their own numbers to wear on the back of their uniforms. Max and Sam are not only on the soccer team, but also on the math team. So, they decide on a very special pair of numbers. When their numbers are squared, they each form a 2-digit square. Furthermore, when they stand next to each other the 4-digit number that is formed is also a perfect square. What numbers did they choose?

A Common Approach

The approach most people attempt is to start with numerals beginning with 1, 2, 3, 4, 5, ... and square each, trying to see which yield a 2-digit perfect square. They then try placing these perfect squares alongside one another to see which will form a perfect square. However, just guessing at random is not the most efficient use of time.

An Exemplary Solution

We can make use of our intelligent-guess-and-test strategy. First of all, we can limit the numbers to select from. In order to form a 2-digit number when squared, the original numbers must be the numbers from 4 to 9, since 1, 2, and 3 when squared yield 1-digit numbers, and 10, 11, ..., 31 squared yield 3-digit numbers. The squares thus formed are 16, 25, 36, 49, 64, 81. Starting with 16, we check to see which pair formed a perfect square when placed side by side. Notice that if we examine 1625 (not a perfect square), we must also examine 2516 (again, not a perfect square). To do the guessing in a clever fashion we would pair the 16 with the remaining possible two digit numbers. When we get to the pair of numbers 16 and 81, when placed side by side give 1681, which is 41^2. The original numbers Max and Sam apparently selected were 4 and 9.

Notice that the numbers 3 and 4 also work, since $3^2 = 9$ and $4^2 = 16$. Standing side by side we obtain 169, which is a perfect square. However, the problem specified they formed a 4-digit square, eliminating this as an answer.

PROBLEM 10.8

Lisa had 26 arithmetic problems to solve for her weekly assignment. In order to encourage her, her dad promised to give her 8¢ for each problem she had correct, but deduct 5¢ for each problem she had wrong. After Lisa finished her assignment, she found out that her dad owed her nothing, and she owed him nothing. How many problems did Lisa get correct?

A Common Approach

A straight algebra approach should enable us to solve this problem.

Let x represent the number of problems Lisa had correct, and y represent the number she had wrong.

Then,

$$8x - 5y = 0,$$
$$x + y = 26.$$

From the first equation, $8x = 5y$ and $x = \frac{5y}{8}$.
Substituting, we get

$$\frac{5y}{8} + y = 26$$
$$5y + 8y = 208$$
$$13y = 208$$
$$y = 16$$
$$x = 10.$$

She had 10 correct and 16 wrong.

An Exemplary Solution

For anyone not familiar with the method for finding the simultaneous solution of two equations in two variables, this problem lends itself to the intelligent-guessing-and-testing strategy. We will keep track of our data in a table. Suppose we start in the middle with 13 correct and 13 wrong.

Number correct × 8¢	Number incorrect × −5¢	Sum
$13 \times 8 = 104$	$13 \times (-5) = -65$	39
$12 \times 8 = 96$	$14 \times (-5) = -70$	26
$11 \times 8 = 88$	$15 \times (-5) = -75$	13
$10 \times 8 = 80$	$16 \times (-5) = -80$	0

Lisa had 10 correct and 16 wrong.

The table of organized guesses easily reveals the answer. Notice that the guesses are not made wildly with no rationale. Rather, we started in the middle and moved up or down, one problem at a time. Since our first guess was way above the answer, we decided to diminish each correct guess by 1, and increase the number of incorrect by 1, diminishing the sum by 13¢ each time.

PROBLEM 10.9

Considering the United States coins of the following denominations: 1¢, 5¢, 10¢, 25¢, 50¢ (not mentioning the 1$ coin). What is the smallest number of coins that will make any sum of money exactly, from 1¢ to $1? What are these coins?

A Common Approach

One approach is to get a number of coins in each denomination and try to find the smallest number needed to make all the sums from 1¢ to $1. In other words, one would then actually perform the required action. Some people will try to work backwards and start with 2 of the 50¢ coins. Neither method is very efficient!

An Exemplary Solution

Use the intelligent-guessing-and-testing strategy. We obviously need four of the 1¢ coins to start so that we can make the values to 4¢. By adding one 5¢ coin we can make every sum from 1¢ to 9¢. Adding a 10¢ coin lets us make all sums up to 19¢. Add another 10¢ coin and we can make all sums to 29¢. One 25¢ coin allows us to make all the sums up to 54¢. Finally, we need one 50¢ coin to complete the set needed to make all values from 1¢ to $1. We need nine coins as follows:

1¢, 1¢, 1¢, 1¢, 5¢, 10¢, 10¢, 25¢, 50¢.

We can test our results by selecting some values at random, and try to make that sum using our nine coins. For example, to make 73¢ we need 50¢, 10¢, 10¢, and 3 of the 1¢ coins.

PROBLEM 10.10

The ancient Egyptians were outstanding mathematicians. The pyramids and the many temples they built attest to this. They were among the first to recognize fractions, and wrote their fractions as the sum of unit fractions. (A unit fraction is a fraction whose numerator is 1.) For example,

$$\frac{5}{6} \text{ would be written as } \frac{1}{2} + \frac{1}{3},$$

$$\frac{3}{10} \text{ would be written as } \frac{1}{5} + \frac{1}{10}$$

$$\frac{11}{18} \text{ would be written as } \frac{1}{3} + \frac{1}{6} + \frac{1}{9}.$$

How would the ancient Egyptians write the fraction $\frac{23}{28}$ as the sum of unit fractions?

A Common Approach

The traditional approach is to list several different unit fractions, find their common denominator and actually add them to find the equivalent set of unit fractions. It would be almost impossible to do this in such a totally disorganized manner. The number of possibilities is almost infinite.

An Exemplary Solution

Just guessing rarely makes sense. Intelligent guessing and testing on the other hand, can be used in an organized manner to resolve this problem. Let us examine the examples given above.

First of all, notice that the denominators of the unit fractions are all factors of the original denominator. In the first example, the denominators 2 and 3 are each factors of the original denominator, 6. Thus, our unit fractions will all have denominators, which are factors of 28. Next, we notice that all the unit fractions start with the largest possible unit fraction, then the next larger one, and continues in this fashion. Obviously the largest unit fraction here, will be $\frac{1}{2}$. Using the factor of 28 as the possible denominators, $\frac{1}{4}$ would be our next unit fraction. If we add these we get: $\frac{1}{2} + \frac{1}{4} = \frac{14}{28} + \frac{7}{28} = \frac{21}{28}$. However, we need $\frac{2}{28} = \frac{1}{14}$ more to get to our desired sum of $\frac{23}{28}$. Therefore, the sought after sum of unit fractions is: $\frac{1}{2} + \frac{1}{4} + \frac{1}{14}$.

The method we used here is for fractions where the denominator is a composite number. A different procedure must be used when the denominator is a prime number.

Index

O

P

Q

R

S

T

V

W

Printed in the United States
By Bookmasters